72nd Conference
on Glass Problems

72nd Conference on Glass Problems

*A Collection of Papers Presented at the
72nd Conference on Glass Problems
The Ohio State University, Columbus, Ohio
October 18–19, 2011*

Edited by
Charles H. Drummond, III

A John Wiley & Sons, Inc., Publication

Copyright © 2012 by The American Ceramic Society. All rights reserved.

Published by John Wiley & Sons, Inc., Hoboken, New Jersey.
Published simultaneously in Canada.

No part of this publication may be reproduced, stored in a retrieval system, or transmitted in any form
or by any means, electronic, mechanical, photocopying, recording, scanning, or otherwise, except as
permitted under Section 107 or 108 of the 1976 United States Copyright Act, without either the prior
written permission of the Publisher, or authorization through payment of the appropriate per-copy fee to
the Copyright Clearance Center, Inc., 222 Rosewood Drive, Danvers, MA 01923, (978) 750-8400, fax
(978) 750-4470, or on the web at www.copyright.com. Requests to the Publisher for permission should
be addressed to the Permissions Department, John Wiley & Sons, Inc., 111 River Street, Hoboken, NJ
07030, (201) 748-6011, fax (201) 748-6008, or online at http://www.wiley.com/go/permission.

Limit of Liability/Disclaimer of Warranty: While the publisher and author have used their best efforts in
preparing this book, they make no representations or warranties with respect to the accuracy or
completeness of the contents of this book and specifically disclaim any implied warranties of
merchantability or fitness for a particular purpose. No warranty may be created or extended by sales
representatives or written sales materials. The advice and strategies contained herein may not be
suitable for your situation. You should consult with a professional where appropriate. Neither the
publisher nor author shall be liable for any loss of profit or any other commercial damages, including
but not limited to special, incidental, consequential, or other damages.

For general information on our other products and services or for technical support, please contact our
Customer Care Department within the United States at (800) 762-2974, outside the United States at
(317) 572-3993 or fax (317) 572-4002.

Wiley also publishes its books in a variety of electronic formats. Some content that appears in print may
not be available in electronic formats. For more information about Wiley products, visit our web site at
www.wiley.com.

Library of Congress Cataloging-in-Publication Data is available.

ISBN: 978-1-118-20587-7
ISBN: 978-1-118-37710-9 (special edition)
ISSN: 0196-6219

Printed in the United States of America.

10 9 8 7 6 5 4 3 2 1

Contents

PROCESS CONTROL

LEGISLATION, SAFETY, AND EMISSIONS

RECYCLING AND BATCH WETTING

Foreword

The conference was sponsored by the Department of Materials Science and Engineering at The Ohio State University.

The director of the conference was Charles H. Drummond, III, Associate Professor, Department of Materials Science and Engineering, The Ohio State University.

The themes and chairs of the four half-day sessions were as follows:

Glass Melting
Martin H. Goller, Corning Incorporated, Corning, NY, Glenn Neff, Glass Service, Stuart, FL, and Tom Dankert, O-I, Perrysburg, OH

Refractories
Matthew Wheeler, RHI Monofrax, Batavia, OH, and Jack Miles, H. C. Starck, Coldwater, MI

Process Control, Safety and Emissions
Larry McCloskey, Toledo Engineering, Toledo, OH, and Phillip J. Tucker, Johns Manville, Denver, CO

Recycling and Batch Wetting
Warren F. Curtis, PPG Industries, Pittsburgh, PA, and Elmer Sperry, Libbey Glass, Toledo, OH

Preface

In the tradition of previous conferences, started in 1934 at the University of Illinois, the papers presented at the 72nd Annual Conference on Glass Problems have been collected and published as the 2011 edition of The Collected Papers.

The manuscripts are reproduced as furnished by the authors, but were reviewed prior to presentation by the respective session chairs. Their assistance is greatly appreciated. C. H. Drummond did minor editing with further editing by the American Ceramic Society. The Ohio State University is not responsible for the statements and opinions expressed in this publication.

This is my final year as Director of the Annual Conference on Glass Problems. I became Director in 1976 and have served as Director for all Conferences at The Ohio State University since that time. I would like to express my sincere appreciation to all the many authors who have made presentations over these years. Without the assistance of the various Advisory Board members the meeting would not be the success that it has been. Their hard work for which they volunteered their time and devotion to have a high quality program has been invaluable.

CHARLES H. DRUMMOND, III
Cocoa Beach, FL
December 2011

Acknowledgments

It is a pleasure to acknowledge the assistance and advice provided by the members of Program Advisory Committee in reviewing the presentations and the planning of the program:

Warren F. Curtis–*PPG Industries*
Tom Dankert–*O-I*
Martin H. Goller–*Corning Incorporated*
Larry McCloskey–*Toledo Engineering*
Jack Miles–*H. C. Starck*
Glenn Neff–*Glass Service*
Elmer Sperry–*Libbey Glass*
Phillip J. Tucker–*Johns Manville*
Matthew Wheeler–*RHI Monofrax*

Glass Melting

OPTIMIZATION OF BURNERS IN OXYGEN-GAS FIRED GLASS FURNACE

Marco van Kersbergen[1], Ruud Beerkens[1], Wladimir Sarmiento-Darkin[2], Hisashi Kobayashi[2],

[1] TNO Glass Group, Eindhoven, NL

[2] PRAXAIR Inc. Tonawanda and Danbury, New York, USA

ABSTRACT

The energy efficiency performance, production stability and emissions of oxygen-fired glass furnaces are influenced by the type of burner, burner nozzle sizes, burner positions, burner settings, oxygen-gas ratios and the fuel distribution among all the burners. These parameters have been optimized for a 300 tpd (metric tons per day glass pull) oxygen-gas fired container glass furnace in order to improve the heat flux into the glass and to improve convective flows in the glass melting tank. An objective of these adaptations was a reduction of foam formation and decrease of evaporation from the melt. Energy consumption and emission of NOx and dust were measured before, during and after the modifications in these combustion parameters. Glass quality parameters such as seed count and glass color were monitored daily and were kept in very tight ranges. The paper describes the combination of burner optimization measures that have been applied and the results concerning energy consumption reductions, emission changes and stabilizing glass quality.

INTRODUCTION

Energy efficiency is a key parameter in the operation of a glass furnace and a main contributor for reducing melting costs. Optimization in energy consumption is always desired and sought. Unfortunately, almost any optimization brings additional capital expenditures that in many cases are difficult or impossible to justify on fuel savings alone. Ideal case would be to optimize furnace performance by adjusting operational variables of existing equipment. By utilizing the tools at hand, benefits can be achieved with minimum added costs. The following sections present the results obtained from a rigorous optimization process carried out, in an European industrial container glass furnace, to improve energy consumption without incurring in major capital expenditures.

OBJECTIVES

The following general objectives were established for the project.

- Decrease foaming and improve heat transfer to melt;
- Increase energy efficiency of glass furnace;
- Stabilize glass quality: fining, color stability, redox stability;
- Keep low emissions (NOx) and low alkali evaporation from melt;
- Optimize burner settings
  - Increase nozzle size (lower gas velocities);
  - Optimize oxygen-natural gas ratio per burner;
  - Change burner orientation;
  - Optimize fuel (natural gas) distribution among burners;

3

FURNACE DESCRIPTION

The glass furnace used during these tests produces green container glass, typically with about 65%- 85 % cullet in the batch (i.e., 65 to 85 % of glass produced from recycled cullet). The typical pull is about 300 to 320 metric tons molten glass per day. This furnace underwent a major cold repair in the summer of 2010. Figure 1 shows a schematic view of the flame coverage in the furnace when fired with ten burners. The numbers indicate the burner locations in the sidewalls.

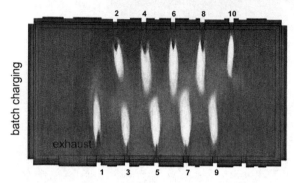

Figure 1. Test Furnace showing flame coverage and burner location

The furnace has eleven Praxair's DOC-JL burners[1] installed. During the tests burner 11 was taken out of operation. These burners are capable of generating a fuel rich luminous flame while keeping very low NOx emissions. Some of the most relevant characteristics of the burners are listed below.

- Deep staging to produce a rich, luminous, "low NOx" flame;
- Oxygen staging to control:
  - o Flame shape, length;
  - o flame stability;
  - o Atmosphere stratification (above melt surface).
- Multi-fuel burner;
- Alkali volatilization control;
- Produce a "targeted" flame.

Figure 2 shows front and rear view of the burner. The furnace also uses Praxair's Tall Crown Furnace design that minimizes silica crown corrosion caused by alkali vapors[2]. This technology combined with optimized burner design and positioning was shown to provide excellent service life performance for silica based crowns.

Figure 2. Praxair's DOC-JL burners. In the picture, upper port is $NG+O_2$, while lower port is O_2

BASELINE OPERATION

For the baseline operation eleven DOC-JL burners were used. As shown in Figure 2 the DOC-JL burners consist of two injection ports; the upper port for natural gas and first stage oxygen injection at a controlled ratio (under-stoichiometric), comprising the burner port itself.

The lower port is only for oxygen gas injection to inject the second stage oxygen to achieve oxidized conditions in the combustion atmosphere near the glass melt surface. For the baseline operation all burners were mounted in the standard configuration, i.e., the secondary oxygen injection port positioned below the natural gas/oxygen port for all burners. During January 12-17, 2011, the following parameters were measured for the baseline operation.

- CO_2, NOx and SO_2 concentration in the exhaust, just after the furnace

- O_2 concentration in this exhaust

- Total sodium and potassium vapor pressures in the exhaust and on several locations in the combustion space just above the melt/foam

- Temperatures in exhaust as measured by suction probe thermocouple

- Photographic record to observe foam layer close to bubbling system

Operational variables modifications

TNO, Praxair and the European container glass company performed several modifications to the oxy-fuel burner settings between January and June 2011. Main modifications are summarized below.

- Burners 3 to 10: O_2 injection port placed above NG/O_2 port, i.e., the burners were flipped over vertically to produce CO rich atmosphere near the glass melt surface. Inverting the burners would expose the foam layer, floating on the glass melt in the hot spot area, to a CO rich atmosphere. Increasing CO concentration in foam's close proximity has been demonstrated to deactivate the surfactants that stabilize foam structure[3].

- Stoichiometric ratio adjusted and fuel input distribution optimized. Oxygen excess was decreased and fuel input distribution changed for burners 3 to 9 to optimize energy consumption by steeper crown temperature profiles to create better flow patterns in the melting tank.

- Burner nozzles were replaced to reduce gas injection velocities by 30% for natural gas and 25% for oxygen. (Higher injection velocities were used in the original installation to try to minimize NOx emissions.)

- Burners 1 and 2 were left operating with the oxygen injection at the lower port to keep an oxidizing atmosphere above batch blanket (to avoid the early decomposition of sulfate fining agent).
- Burner 11 was shut down.

These changes were made with the following technical considerations. First, it appeared rather difficult to apply a low oxygen excess on all burners and still achieving complete combustion and low CO levels in the exhaust gas flow plus good glass quality. Therefore, the two burners closest to the flue port were set on higher oxygen excess to compensate for the total oxygen supply in the furnace. Second, oxidizing conditions above the batch are important for sulphate retention and batch redox as well as to prevent high SO_2 emissions.

Adjustments in the operational conditions were made in six steps, named A, B, C, D, E and F. In this paper details concerning each step are not fully explained and only the results after the last phase (F) will be compared with those from the baseline operation. The following figure shows the changes performed in the fuel distribution of the furnace.

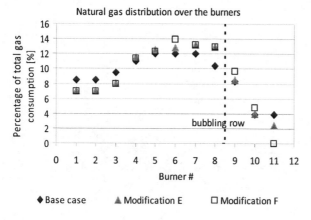

Figure 3. Fuel distribution in each burner

As can be seen in Figure 3, the fuel distribution was changed to favor a higher energy delivery in the hot spot zone of the furnace (creating a more distinct hotspot). Last burner (B11) was shut down to compensate for the additional temperature given to the glass in the hot spot. Figure 4 shows oxygen fuel ratio optimization during the test. Changes performed also contributed to stabilize furnace operation (such as crown temperatures), as shown in Figures 5 and 6.

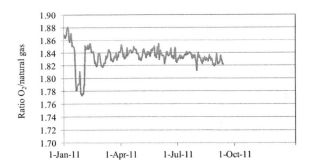

Figure 4. Total Oxygen/Natural Gas ratio for combustion process during the test

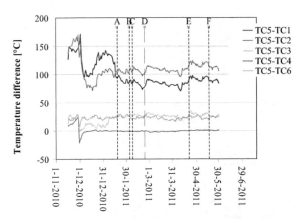

Figure 5. Crown temperature difference relative to the hot spot (T5)

Figure 5 shows crown temperatures at different crown thermocouple locations with respect the hot spot (TC5). Magnitude of the difference as well as the variations experienced over time decreased after modification A. Similar behavior is described in Figure 6.

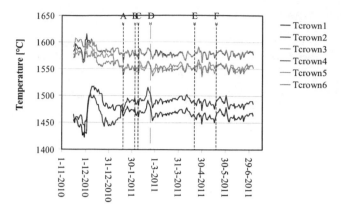

Figure 6. Furnace crown temperatures

Impact of the modifications performed in the operational variables on foam characteristics was difficult to quantify by observation only. Pictures in Figure 7 show foam layers near the furnace bubbling area before and after the modifications. The burner modifications appear to have reduced the thickness of the foam as can be seen in the areas circled in dotted lines in Figure 7.

(The visible height of the tuck stone above the foam layer is less in the left picture.)

Preliminary quantification seems to indicate that foam thickness was reduced from about 2 to 1.5 inches. More work is planned in this area to optimize the effect on foam reduction.

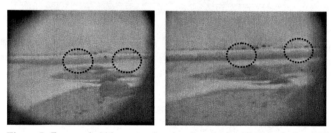

Figure 7. Furnace bubblers area showing foam before (left) and after (right) modifications

Figure 8 shows the impact of the modifications performed on the specific fuel consumption and other changed process conditions. The energy efficiency dropped from about 3.60 MMBtu/ton (3780 MJ/metric ton) to about 3.40 MMBtu HHV/ton (3570 MJ LHV/metric ton) during the test period, which represents a reduction of about 5 to 6%. Not all of the reduction is due to the burner optimization as other factors such as the ambient temperature and seasonal variations in cullet moisture content also contributed.

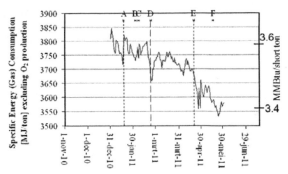

Figure 8. Furnace specific energy input by natural gas in January-May 2011.

Energy consumption increased after the end of the test, but this increase was driven by a decrease in the level of cullet used which went down from 85 to 65%. It is important to point out that this improvement has been made without changing any component or equipment in the plant, just by setting the burners in different ways and using the same furnace, the same glass at the same pull. Glass quality remained similar before, during and after the test as shown in Figures 9 and 10. Quality indicators were within established limits with no major variations resulting from the modification implemented in the furnace operating conditions. Color Dominant Wavelength shown in Figure 9 was stable and always in the required range and seed counts were low (Figure 10).

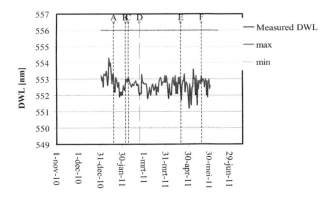

Figure 9. Wave length (glass color) measurement during the test

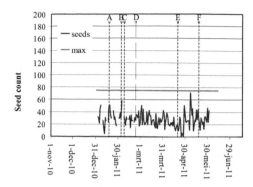

Figure 10. Seeds count measured during the test

Table I. The following table shows a summary of measurements in the flue gas exhaust port.

Measured concentrations in dried exhaust gases. All concentrations are calculated to normal conditions (dry, 1 bar, 273 K, indicated as Nm^3)

	O_2 [vol.-%]	CO_2 [vol.-%]	CO [vol.ppm]	SO_x [mg/m_n^3]	NO_x [mg/m_n^3]	NO_x [kg/ton glass]
13-Jan	3.2	70	0	2735	1601	0.24
18 Jan				Modification A		
24-Jan	5.6	67.7	>1000**	2884	1986	0.30
3 Feb				Modification B		
7-Feb				Modification C		
23-Feb				Modification D		
18-March	7.8	70	400	2652	2419	0.36
20 -Apr				Modification E		
18-May				Modification F		
26-May	3.4	73.8	500	3700	1318	0.20

First row of the table (Jan 13th) shows values selected as baseline. Measurements were performed by TNO at the same location in the flue gas channel, close to the exit of the flue gases from the furnace. Gas flow was extracted using a water cooled suction probe, then all water was condensed and the dried extracted gas was analyzed on CO, O_2, CO_2, NOx, SO_2. The concentrations are based on the contents in the dried flue gas at actual O_2 content. SOx concentrations are SO_2 and SO_3 content together expressed in SO_2 and NOx concentration is the total contents of nitrogen oxides (mainly NO and NO_2) expressed as NOx. Although CO measured concentrations at furnace exit increased slightly to 500 ppm, CO values reduces back to almost 0 in stack. The furnace is equipped with a bag filter and a scrubber so they have almost no dust emissions.

NOx levels decreased about 17% after performing all modifications from 0.24 kg/Mton (0.48 lb/ton) to 0.20 kg/Mton (0.40 lb/ton). Even though the natural gas used at the plant contains about 14% nitrogen in its composition, NOx performance of Praxair's DOC-JL burners was shown to be extremely low compared to typical NOx emissions from commercial glass furnaces. In Figure 11, NOx emissions measured in this furnace were compared with data from other oxy-fuel fired container glass furnaces. Three key parameters influencing the NOx emission from oxy-natural gas fired furnaces are N_2 concentration in the furnace, burner design and niter in batch. In general NOx emission from an oxy-fuel burner increases proportionally to the nitrogen concentration (measured in volume % on a wet basis) in the furnace. Conventional oxy-fuel burners with high flame temperature typically produce 1.0 to 2.0 lb/ton of NOx in a well sealed furnace with a N_2 concentration of 5-10 %. NOx emissions from low flame temperature oxy-fuel burners, such as Praxair's DOC-JL Burner, are reduced to roughly 1/5th to a range of 0.2 to 0.4 lb/ton (0.1-0.2 kg per metric ton) under the same N_2 concentration in the furnace. Data collected during the test were higher than this range as the natural gas contained 14% volume N2. It is known from laboratory burner tests that N_2 contained in natural gas produces higher NOx emission than N_2 contained in the oxygen supplied to the burner and N_2 in the infiltrated air when the furnace N_2 concentration is kept constant[4,5].

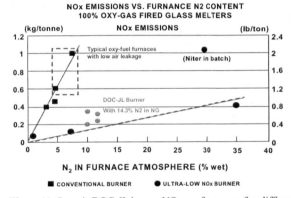

Figure 11. Praxair DOC-JL burner NOx performance for different N_2 in-furnace concentration

Particles and vapors measured during the tests are summarized in Table 2. No major variations were observed during the test indicating that modifications performed did not altered substantially critical variables acting in volatilization of glass components.

In Figure 12 the range of alkali vapor (NaOH+KOH) values measured through the furnace view ports during the test is shown with the corresponding crown temperature range. The alkali vapor concentration profile in the furnace depends on the furnace gas recirculation pattern and is generally higher at the glassmelt surface and lower near the crown[6]. Since the measured values are at the elevation of the furnace view port (about 20 cm above the glass melt surface), the alkali vapor concentration near the crown was estimated by extrapolation from previous data from CFD analyses and measurements from other oxy-fuel fired furnaces. The figure also shows the safe and critical ranges of furnace operation to control corrosion of silica crown by alkali vapor attack for both conventional silica and no lime silica (i.e., silica without calcium binder). The silica corrosion criteria were developed from extensive experimental and theoretical studies at TNO[7-8]. When a furnace is operating in the critical zone it has much higher probabilities of suffering from corrosion attack/damage than a furnace operating in the safe range. Alkali vapor values measured in this furnace fall comfortably in the safe range as the furnace uses no lime silica crown.

Table II. Particles and vapors measured during the test

Values in Pascal

	Base	After A	After D	After F
Al_2O_3	0.014	0.019	0.019	0.020
As_2O_3	0.044	0.052	0.047	0.064
HBO_2	2.466	2.288	2.239	2.700
CaO	0.802	0.145	0.137	0.122
CdO	0.004	0.007	0.008	0.012
Cr_2O_3	0.654	0.350	0.363	0.241
CuO	0.013	0.005	0.002	0.004
KOH	1.065	1.025	1.016	1.136
MgO	0.325	0.085	0.053	0.062
NaOH	14.985	15.242	15.200	15.803
PbO	0.093	0.141	0.132	0.207
SO_2	34.907	36.817	33.864	47.228
Sb_2O_3	0.008	0.007	0.006	0.008
SeO_2	0.060	0.057	0.062	0.084
SiO_2	0.341	0.321	0.285	0.302
SnO	0.035	0.039	0.035	0.025
V_2O_5	0.011	0.010	0.010	0.002
ZnO	0.006	0.051	0.033	0.069

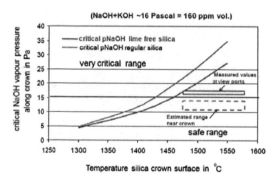

Figure 12. Alkali concentration vs. temperature - Critical operational ranges for Silica crowns

CONCLUSIONS

Through optimization of oxy-fuel burner settings and firing conditions the natural gas consumption was reduced by 5.8 % and the oxygen consumption by 3.8 % during and after the test period. Since the contribution of lower foaming is expected to influence energy consumption by maximum 2.0 %, the good test results were also contributed by the effects of other changes in operation during the test period (January 2011- June 2011) such as ambient temperature, batch/cullet humidity, oxygen excess etcetera. The following table summarizes expected ranges of contributions from different effects.

Table III - Expected ranges of contributions from different effects.

Changed parameter	Estimated effect on energy consumption
Moisture content in the batch (+ cullet) (2-5 % moisture)	2.5-3.5 %
Ambient temperature (4-17 °C)	0.5 - 2 %
Reduced foam height by Burner adjustment	1.5 - 2.0 %
Oxygen/Gas ratio 1.88 → 1.84	0.5 %

Sodium and potassium vapor pressures did not increase with burner adjustments and were less than 17 Pascal in all measurements. No risk of increased silica corrosion is expected after the modifications. NOx emissions were reduced about 17% from 0.24 kg/Mton (0.48 lb/ton) to 0.20 kg/Mton (0.40 lb/ton) – Mton = metric ton. Glass quality was kept stable during and after the test period. Fluctuations in seeds count and DWL (dominant wavelength determining glass color) appeared to have become less frequent and glass quality was good. Exhaust temperatures measured by the suction probe pyrometer were stable during the test. A small decrease in the furnace exhaust temperature of about 5-10 °C was measured, indicating improved heat transfer to the melt.

ACKNOWLEDGMENT

The authors wish to acknowledge the contributions of Agentschap NL and Mr. Léon Wijshoff for the support to this project.

REFERENCES
1. U.S. Patents 5,266,025 , 5,411,395, 5,267,850 , 5,295,816, 5,755,818, 5,924,858 , 6,394,790
2. U.S. Patent 6,253,578
3. Laimböck, P.R., "Foaming of Glass Melts", Ph.D. Thesis, Technical University of Eindhoven, 1998
4. Kobayashi, H., G. B. Tuson and E. J. Lauwers, "NO$_x$ Emissions From Oxy-Fuel Fired Glass Melting Furnaces," European Society of Glass Science and Technology Conference on Fundamentals of the Glass Manufacturing Process, Sheffield, England, September 9-11, 1991.
5. Kobayashi, H., Beerkens, R. G. C., Errol, P., and Barbiero, R., "Emissions of Particulates and NOx from Oxy-Fuel Fired Glass Furnaces," European Society of Glass Conference, Venice, Italy, June 21 to 24, 1993.
6. Wu, K. T. and H. Kobayashi, "Three Dimensional Modeling of Alkali Volatilization / Crown Corrosion in Oxy-Fuel Fired Glass Furnaces", The 98[th] Annual Meeting of the American Ceramic Society, Indianapolis, IN, April14-17, 1996
7. Faber, A.J. and R.G.C. Beerkens, "Reduction of Refractory Corrosion in Oxy-Fuel Fired Glass Furnaces", in Proceedings of the XVIII Internati[on]al Congress in Glass, 1998
8. Beerkens, R.G.C. and Verheijen, O.S., "Reactions of Alkali Vapors with Silica Based Refractory in Glass Furnaces, Thermodynamics and Mass Transfer. Phys. Chem. Glass, (2005) vol. 46 no. 6

FUTURE ENERGY-EFFICIENT AND LOW-EMISSIONS GLASS MELTING PROCESSES

Ruud Beekens, Hans Van Limpt, Adriaan Lankhorst, and Piet van Santen
TNO Glass Group, Eindhoven, The Netherlands

ABSTRACT

All over the world, there is an increasing drive to develop new technologies or concepts for industrial glass melting furnaces, with the main aim to increase the energy efficiency, tabilize production and reduce emissions. The application of new process sensors, improved furnace design, intelligent control strategy and waste gas heat recovery systems support the glass manufacturers to achieve ever increasing requirements concerning energy efficiency and emission limits. Glass industries are searching for breakthrough innovations (revolution), but introduction of such major changes in industry is hampered by large risks and investments. Therefore most companies prefer a stepwise improvement (evolution) based on existing concepts, such a regenerative glass furnaces. When different technologies are combined, it will even be possible, in specific cases, to avoid the use of flue gas scrubbers or DeNOx systems. An overview of requirements for industrial glass furnaces is given concerning melting characteristics and performance of the furnaces. A short overview of evolutionary and revolutionary developments in industrial glass melting processes will be presented. In this paper, an example of an optimized, regenerative end-port fired glass furnace concept, based on a combination of current (BAT) technologies, will be given in more detail. Basic innovative elements of such a furnace will be shown and its expected performance.

1. INTRODUCTION

1.1 Elementary steps in industrial glass melting

The essential steps in the sequence of the process of glass melting are [1, 2]:

- Melting-in of primary raw materials by several chemical reactions and (eutectic) melting processes, involving the release of gases and vapors. This process, to bring the raw materials to reaction temperatures and to provide the energy for endothermic processes, requires the main part of the heat transferred to the melting tank;

- Reactive dissolution of sand grains in primary melts: important are the presence of fluxes in the batch blanket and the sand grain size distribution. This process takes place at the borders of the batch blankets. The melting kinetics are strongly determined by formation of eutectic melts, that may segregate from the sand due to their low viscosity. Kim et al. [3] suggest new selective batching methods (based on combining premixes of two or three types of raw materials from the batch) to avoid the de-mixing in the batch blanket and to improve the kinetics of silicate glass forming reactions;

- Dissolution of residual sand grains in the freshly molten glass, supported by convection (stirring) of the melt. Dissolution of sand grains should take place below the fining onset temperature to avoid overlap of sand dissolution and fining;

- Removal of small seeds, often containing remaining batch gases (e.g. air gases and mainly CO_2) by the primary fining process, taking place above the fining onset temperature [1]. At this temperature the primary fining process is enhanced by the start of the decomposition of the fining agent. The fining onset temperature needs to be achieved for all glass melt flows in the melting tank to obtain a melt without small seeds. The fining onset temperature is mainly determined by the batch composition and more specific the content of fining agent, type of fining agent, water vapor pressure in the furnace atmosphere and presence of cokes (or other redox-active species) in the batch. In the primary fining process, small seeds are exposed to gases released from the melt by fining agents and this will result in bubble growth and increased bubble ascension to the glass melt surface. A shallow melting tank or fining section will favor the removal of gas inclusions from the melt;

- Re-absorption of remaining seeds, after the primary fining process. Seeds that contain gases which dissolve chemically in the silicate melt might be digested by the melt (this is the case for gases with increasing solubility in the melt at decreasing temperature) during conditioning and controlled cooling of the glass;
- Homogenization of the melt: for optical glasses or display glasses, the chemical homogeneity is utmost important and can be improved by dynamic stirrers in special spaces (stirring chambers), in the forehearth or feeder systems. For other glass types, thermal uniformity is more important. Static stirrer systems have been hardly used for glass melt conditioning or homogenization, but may be of future interest.
- Conditioning of the melt, prior to forming. Conditioning brings the glass melt to a uniform or at least controlled viscosity level, suitable for the applied forming processes.

1.2   The ideal glass melting process

Figure 1 shows diagrams with the main process steps in industrial glass production and their optimum conditions for the case of soda-lime-silica glass [1], starting from batch preparation and as an energy efficient option batch-preheating, then batch blanket fusion, sand grain dissolution, primary fining, secondary fining and conditioning. The optimum conditions for each project step is roughly indicated.

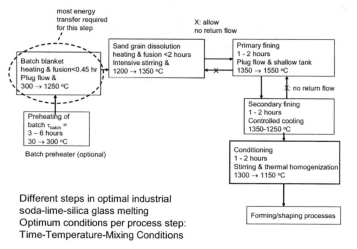

Different steps in optimal industrial soda-lime-silica glass melting
Optimum conditions per process step:
Time-Temperature-Mixing Conditions

Figure 1   Industrial glass melt process split in subsequent process steps, ideal conditions for each process step are indicated. Heating of batch blanket requires 80-85 % of total heat input and residence time in batch blanket is only 2.5-5 % (45-75 minutes) of total average residence time.

The return flow from the fining section into the melting-in section or from the conditioning section back into fining zone is often very strong and the first return flow is required for heat transport to the batch blanket tip. Most recirculation flows are just caused by temperature differences (free convection) in the melt. However, these return flows lower the melting efficiency (space utilization [4]) of the melting tank and causes very wide residence time distributions. In today's glass furnace, the return flows are hardly controlled, due to limited understanding of the glass melt flow patterns and the limited adjustable process settings, that influence the flow patterns in the melt (bubbling, barrier boost settings, hot spot adjustments).

Apart from these desired process steps, non-desired phenomena may take place such as glass melt foaming, NOx formation in the combustion space, evaporation of volatile glass melt components (leading to emissions, e.g. of dust or metals or attack of refractory by the volatiles), refractory corrosion in the melting tank or superstructure.

Therefore, it is not only important to design for each of above mentioned steps a dedicated section, but it is also important to select process conditions that limit energy losses and that minimize emissions or to avoid circumstances, that may spoil the refractory material. Today, most tank furnaces have a rather simple one-tank design, but for optimizing each process step, a segmented melter is considered to be more suitable and enabling a more compact design [1]: lower tank volume to pull ratio. Different techniques can be used for the different segments, for instance in the primary fining zone different methods may be applied to speed up fining, like: centrifugal systems, low pressure fining, application of barrier boost, application of a fining shelf or bank. In this paper, conventional glass furnace concepts are the starting point.

1.3   Overview of current developments in glass melting technology: evolutionary approaches and on the long term: revolutionary approaches.

Table 1 shows a short, but rather comprehensive overview of current developments in modern furnace design aspects and more innovative new technologies for glass melting. A large number of patents and publications can be found on these design aspects.

Barton [5] presented an overview of different melting concepts for glass in the 1990-ties, new developments since then, but with limited application sofar are: submerged combustion melter, segmented melters (e.g. for optical glasses), centrifugal fining, low-pressure fining, plasma-melt processes and to a larger extent, improved oxygen-fired furnaces.

Still many furnaces are based on regenerative firing, but stepwise improvements for such furnaces have been applied in electric boosting systems, combustion space and burner port designs, refractory materials for crowns and checkers and controllable bubbling systems plus batch preheating devices. Furthermore, a development towards advanced process control (multiple input – multiple output, feed forward, process model based) [6], application of redox (pO_2) sensors [7] for the melt or tin bath [8] and combustion control using sensors for CO and O_2 monitoring [9] in the exhaust gas can be observed.

Table 1   Overview of innovative developments in glass furnace designs

Technology Name	Lit. no	Specific details/description	Targets / Expectations			
			NOx	Energy	Glass quality	Specific Pull or Average Residence
EVOLUTION						
End-port fired LowNOx design	13	Optimized combustion space and burner port design	< 600 mg/Nm³ < 1 kg/ton	top 10 % most energy efficient	normal container glass	3-3,5 ton/m² day
Cross-fired regenerative LowNOx design	13	Optimized combustion space and burner port design	< 600 mg/Nm³ < 1 kg/ton	top 10 % most energy efficient	normal container glass	3-3,5 ton/m² day
		Optimized combustion space and burner port design (e.g. further development FENIX)	< 850 mg/Nm³ < 2 -2.5 kg/ton	top 10 % most energy efficient	float glass	2,25 ton/m² day
High Pull Oxygen furnace (HPO)		Special glass & Container Glass	< 0,3/ton	top 5 % most energy efficient	all glasses	> 4 ton/m² day
Next generation LoNOx melter	19	In combination with oxygen firing and fining shelf	< 0,4 kg/ton	top 10 % most energy efficient	container glass	3-3,5 ton/m² day
Optimized Space Utilization Melter	20	Design configurations with more efficient utilization of space and controlled residence times including improved boosting system	< 600 mg/Nm³ < 1 - 1,5 kg/ton	top 10 % most energy efficient	all glasses	residence time average < 12 hours
Oxygen fired float glass furnace	21	Optimize waste heat recovery and furnace design adapted to oxygen firing	< 1 kg/ton	top 10 % most energy efficient	float glass	2,25 ton/m² day
REVOLUTION						
Segmented melter (TNO)	1	High heat transfer melting-in unit, special finer and refiner	< 600 mg/Nm³ < 1 - 1.5 kg/ton	top 5 % most energy efficient	tableware & cosmetic glass quality	residence time average < 8 hours
Submerged Combustion Melter (SCM)	22	Water cooled walls, bottom burners (oxygen-gas), melt contain foam and seeds	estimates not available	estimates not available		< 2 hours residence time, potential first step segmented melter
Plasmelt/High Intensity Plasma Melting	23	Plasma with argon gas, high electricity consumption, fast melting for high viscous glasses	0	100 % electricity	High melting glasses	Very high specific pull
In-flight Glass melter (NEDO)	24	Batch dosing into vertical flame, collection of melt in refractory tank for fining flame plasma using torches	estimates not available	50% less CO₂ emissions	Normal and high-melting glasses	Very high specific pull, small compact systems
Thin-layer melter	25	Very thin glass melt layer to increase heat transfer and fining	estimates not available	estimates not available		
Centrifugal finer	26	Relatively slow rotation to enhance bubble removal - part of segmented melter	estimates not available	estimates not available	Special glasses	
Advanced Glass Melter	27	Similar to NEDO in-flight melter	estimates not available	available		
Low-pressure fining / ASAHI Glass	28	Part of segmented melter system	estimates not available	estimates not available	Special glasses	Display glasses

2.   Evolution of regenerative glass furnace design & operation

The furnace design and operation system of a conventional regenerative glass melting furnace has been improved with respect to energy efficiencies and emissions. The design of the furnace is optimized at TNO by the application of CFD simulation modelling [10] of the glass melting & combustion process in a regenerative glass furnace, and it constitutes a combination of the following features:
-   Optimized insulation of the furnace walls, crown and bottom, using CFD modelling to allow high level of insulation, without jeopardizing furnace lifetime;
-   Increased combustion chamber and adapted burner port dimensions to create conditions of delayed mixing of fuel and air, combustion gas re-circulation (dilution of the root of the flame) and still complete combustion. This will result in high energy efficiency (long residence times of combustion gases in combustion space) and decreased NOx levels;
-   Optimize positions of burner ports, number of burners and types of burners;
-   Apply advanced control systems to optimize the combustion conditions, using O_2 and CO sensors in the exhaust gas flow [9] and use model based control strategies [6] for stabilizing flow patterns and temperatures of the glass melt in order to operate at optimum energy efficiency and glass quality conditions. Glass melt quality aspects are often linked to stable and optimum flow conditions (avoid short cut flows) in the melting tank and control of the position of the spring zone and hot-spot to achieve optimum fining also at high pull rates. Today advanced CFD-model based control systems are able to determine the hot spot as well as the spring zone and to stabilize the positions of these zones;
-   An overall on-line energy balance calculation is the base for stabilizing and optimizing the energy supply (boosting and fuel) to the furnace. **Figure 2** shows the energy distribution of a non-optimized end-port fired glass furnace, **the figure also** shows the energy split and energy consumption for a modern end-port fired furnace with efficient regenerators, highly insulated, tight combustion control and one example of an optimized furnace with batch/cullet preheating [11];

* *CFD = Computational Fluid Dynamics models are based on the basic laws of mass, energy, momentum and electric charge conservation. Application of these models will provide information on local temperatures, heat transfer rates, flow patterns in melt and combustion space and in combination with mass transfer, thermodynamic or thermo-chemical or reaction-kinetics models give an estimation for sand grain dissolution rates and fining performance.*

-   Application of economic feasible batch & cullet pre-heater systems, in combination with adapted doghouse design [12] or optimized burner positioning with low flame velocities, to recover flue gas energy, but without increasing carry-over tendency by charging dry batch into the melting tank. Direct contact between the flue gases and batch will offer the additional functionality of a gas scrubbing system: only a filtration system for the removal of batch and primary glass furnace dust is required, but no DeNOx or hardly any additional scrubbing systems are necessary for a well designed furnace with combustion and glass melt control systems plus batch preheating.

Computation Fluid Dynamics (CFD) modelling [10], to simulate the glass melting process in the melting tank as well as simulation of the combustion process and evaporation processes [13], has been used to optimize this furnace design. **Figure 3** shows as an example, a result of the CFD-modeling of the end-port furnace combustion space (flame temperature contours). The optimized furnace is well insulated and heat losses through the bottom and crown are about 1 to 2 kW/m^2, sidewalls are efficiently insulated, where it is possible, with an average sidewall heat loss of ± 3 kW/m^2.

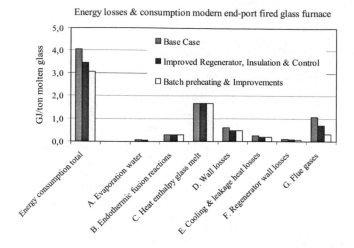

Figure 2.   Energy distribution in standard end-port fired furnace and optimized end-port fired furnace (improved regenerators, process control and optimal insulation) and third option with batch preheating. Furnace with 50 % cullet, 2.9 tons/m^2.day

The regenerator is designed to have an efficiency of about 70 % (sensible flue gas heat transferred to combustion air).

The optimized furnace has a relatively large combustion chamber and spacious burner ports. **Figure 4** shows the essential distances in an end-port fired glass furnace that are important for lowering NOx formation and optimizing energy efficiency. The vertical angle of the burner ports and the cross section area are optimized to create long and luminous flames with low NOx formation tendency. Typically, 2 nozzle-adjustable lowNOx burners are installed at each side. Decreased NOx formation and an optimal heat transfer of the flames to the melt are realized, applying controlled and slow mixing of combustion air and fuel and nearly stoichiometric combustion conditions, controlled by flue gas analysis in the burner port by continuous O$_2$ and CO analysis.

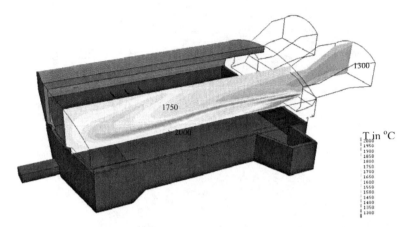

Figure 3   Example of a result of the CFD modelling of the processes in the combustion space of an end-port fired glass furnace. The temperature contours show the temperatures in a cross section of the U-flame.

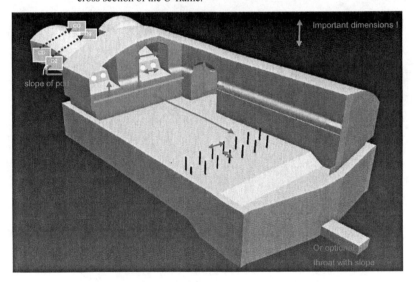

Figure 4   Sketch of the interior of an end-port fired furnace with some <u>essential distances</u> and <u>dimensions important</u> with respect to energy efficiency and/or NOx emissions. Burner port with the trajectories of laser beams for O_2 and CO measurements is indicated by dashed lines. The most optimum dimensions and distances can be derived from CFD modelling studies and are to be determined on a case-by-case analysis.

In the relative large combustion chamber of this end-port fired regenerative furnace, the hot combustion gases are partially re-circulated into the core of the flame, the maximum temperatures and oxygen content of the flame are reduced and the flames are sooty (carbon soot with high emissivity) and luminous. Post combustion in the top of the regenerator is avoided. Reducing flame parts will not touch the batch blanket and premature dissociation of sulphates in the batch blanket is avoided.

Figure 5 shows the achieved CO (top regenerator) and NOx emission levels for a conventional end-port fired container glass furnace equipped with a CO and oxygen sensor and controlling combustion by air-natural gas ratio. This figure shows the importance of tight combustion air:fuel ratio control based on CO/O_2 analysis in the exhaust gases.

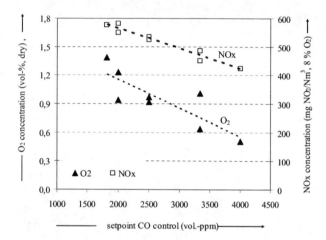

Figure 5   NOx emissions of end-port fired container glass furnace, using CO and O_2 sensors in burner port and combustion air control

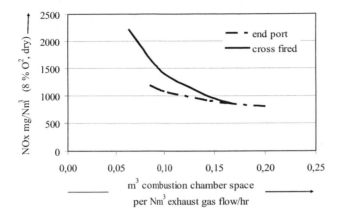

Figure 6   Effect of ratio combustion space volume / gas consumption on NOx emission levels of a
            set of end-port fired regenerative furnaces.

Figure 6 shows the effect (from benchmark analysis) of the relative volume of the combustion space
(with respect to natural gas consumption) and the NOx emissions. A large combustion space, generally
shows lower NOx emission levels.

3. NO_X AND ENERGY BENCHMARKING OF REGENERATIVE GLASS FURNACES
3.1   NOx benchmarking

For a set of about 50 regenerative glass furnaces, a NOx benchmark study has been performed in 2008.
The results have been used to establish statistically found relations between furnace design aspects and
process parameters at one side and NOx formation at the other side. The NOx emissions of the
investigated end-port glass furnaces vary roughly between 500 and 2000 mg/Nm^3. In general, oil
flames produce less NOx than natural gas flames, because in general the oil flames contain more soot
and flames cool down more rapidly. The statistical correlations obtained from the NOx benchmark
study among others show that air excess, vertical burner angle, nozzle size, number of burners per port
and the volume of the combustion chamber (versus fuel consumption) are the most important criteria
for a glass furnace design with low NOx formation levels.

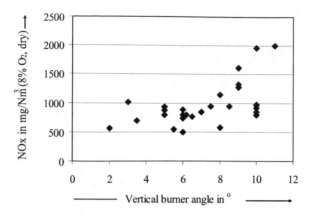

Figure 7 Impact of vertical burner angles on NOx formation in end-port furnaces. The NOx concentrations are given in mg/Nm³ (T = 0 °C, p =101.3 kPa), measured at dry flue gas conditions and values normalized at 8 % O_2.

No correlations were found between NOx emissions and pull rate of a furnace, neither on the N_2 contents in natural gas (the N_2 contents in natural gas seem only to have an noticeable effect on NOx emissions of oxygen gas fired furnaces) and age of the furnace seem to have significant influence. As an example, **figure 7** shows the impact of vertical burner angles (angle between burner axis and horizontal) on NOx formation in end-port furnaces.

3.2   Energy efficiency benchmarking
Glass furnace energy benchmarking studies **[14, 15]** are used to rank the energy efficiency of glass furnaces in the same glass industry sector (e.g. container glass, float or fibre glass). The goal of benchmarking is mainly the comparison of energy efficiency or specific energy consumption of glass furnaces (or feeders) within a glass sector and to identify the most energy efficient option: "best practice". **Figure 8** shows the effect of type of glass and pull on specific energy consumption of glass furnaces according to benchmark analysis of industrial glass furnaces between 1999-2007.

Energy Consumption of Glass Furnaces in Container glass sector
In the benchmark database, the container glass furnace with the highest energy efficiency shows a specific (annual averaged) primary energy consumption of 3.8 GJ/ton molten glass at a level of 50 % cullet in the batch. Compared to other reported values in literature, this seems to be rather high, but in this case the energy consumption through electricity is calculated as primary energy equivalent (1 kWh = 9 MJ) assuming an overall power plant efficiency (including electricity transport) of 40 % instead of the net value (1 kWh = 3.6 MJ) and the data are normalized to a level of 50 % cullet in the batch.

The identified **best practice** of an existing energy efficient container glass furnace is an end-port fired furnace, with regenerator designs and structures resulting in regenerator efficiency above 65 % (more than 65 % of the sensible flue gas heat is transferred to the combustion air) and with cullet/batch preheating up to temperatures of 275-325 °C, only moderate electric boosting, high level of insulation of the crown, and optimum sealing of all joints.

The furnace typically operates at a pull above 250-300 tons per day with a specific pull of ±3 metric tons/m^2 per day.

The burners and tight combustion control leads to luminescent flames with a moderate excess of air (0.8-1.2 vol. % of oxygen in the exhaust gases), but CO contents in the exhaust gases should be limited.

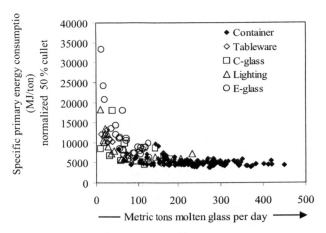

Figure 8   Effect of type of glass product and pull on annual average specific energy consumption of glass furnaces, according to benchmark analysis of industrial glass furnaces. Residence times of the melt in the furnaces are very different for different furnaces & glass types. About 200 industrial glass furnaces included.

Conradt [16] showed that, taking into account the driving force required for heat transfer and limited time available for heat transfer and the laws of thermodynamics, the minimum practical possible energy consumption calculated for a fossil-fuel fired container glass furnace with very large regenerator and being optimally insulated is about 3.30 GJ/ton glass melt for 70 % cullet in the batch (without batch preheating), and normalized to 50 % cullet, this minimum practically achievable energy consumption will be about 3.5 GJ/ton.

Thus the difference between the energy consumption of the most efficient container glass furnaces, today (3.6-4.2 GJ/ton molten glass from batch with 50 % cullet) and the minimum possible energy consumption is already very small.

The existing furnace with the lowest energy consumption, not taking into account the primary energy required for oxygen production and electricity production (1 kWh = 3.6 MJ) and not normalizing to 50 % cullet was an oxygen-fired furnace, 60 % cullet, using on average 3380 MJ energy per ton molten glass. However normalization to 50 % cullet and conversion to primary energy equivalent (including electricity use for oxygen supply) [14] will result in a much higher value for this case: about 4200 MJ/ton. This example shows the importance of such normalization or correction rules, having a large impact on the ranking of the energy efficiencies. Without such normalization, all-electric melters would give the best performance, although the furnaces often consume indirectly a lot of primary energy.

Table 2 shows some measures and their typical impact (% energy savings) on the energy consumption of an end-port fired container glass furnace (320 tons/day, base case: 4.19 GJ/ton glass melt)

Table 2   Several measures and indicated energy saving potential by these individual measures starting from base case end-port fired glass furnace

Measures	320 tons glass melt/day	
	MJ/ton	% savings
Base case	4.192	0
Batch humidity 3.5 → 2 %	4.08	2.7
Emissivity flames 0.18→ 0.25	4.13	1.4
20 % better insulation	3.98	5.2
Batch preheating 300 °C	3.43	18
Air excess 10 % → 5 %	4.15	1
Cullet 40 → 75 %	3.81	9
Crown 10 % higher	4.22	-0.6
Regenerator 63 → 68 %	3.991	4.8
Throat temperature 1325→ 1300 °C	4.105	2
No cold air infiltration -500 Nm^3/hr	4.12	1.7

4.   IMPORTANT DESIGN AND OPERATIONAL ASPECTS OF GLASS FURNACE WITH LOW NOX EMISSIONS AND INTENSIVE HEAT TRANSFER

The design of the combustion space is not the only important aspect of a modern, energy efficient regenerative glass melting furnace with low NOx levels.
Basic elements of modern regenerative glass furnaces are:
- Design of combustion space: important aspects and elements are:
  - The slope of the burner port(s), the angle at which the combustion air flows into the combustion space;
  - The size of the burner ports: the preheated combustion air velocity (actual value) should preferably be below 8-9 m/s at the entrance of the combustion chamber;
  - The level (height) of the burner bank above the glass melt line, generally an increased distance will allow more recirculation flows and flame dilution, which results in lower peak flame temperatures and less NOx formation. Furthermore, elevated flames will lower the risk of sulphate decomposition in the batch by flames and will lower carry-over levels;
  - Positions and angle of burners in burner port;
  - Height of combustion chamber;
  - Number of burners and positions of burners in ports;
  - Distance between burner port and doghouse;
  - Dimensions of doghouse;
  - Special cavities in crown to avoid silicate melt run-down along the breastwalls;
  - Tight sealing to prevent cold air infiltration.
  - Adjustable burner nozzles* for flame shaping and controlling the fuel injection velocity levels;
    *Choice of burners to allow adjustable flame shape and flame dimensions and to

- Design of Melting tank, including type and positioning of electric boosting system, electrode connection & arrangements [17];
- Efficient regenerator with 70 % or more energy transfer from available flue gas heat contents to the combustion air;
- Advanced control systems [6] based on CFD type simulation of modelling of the dynamic behaviour of glass melting and combustion processes. Furnace settings are always kept at optimum conditions to enable optimum energy efficiency, high glass melting stability, low glass defect levels and low emissions [13]. The modelling results in model based control systems support the continuous adjustment of input parameters such as fuel input, fuel distribution, air-fuel ratio, power supply on electrodes, batch composition and batch redox in order to achieve optimal melting conditions;
- Control system for combustion process, operating with optimum air-excess and controlling NOx and CO emissions
  The combustion process and combustion air flow can be controlled with modern in-line laser CO sensors in both burner-ports [9].
- The furnace is equipped with the latest generation batch/cullet or pellet preheating systems to recover waste gas heat.
  In case of a pre-heater of the type with direct contact between the flue gases and the batch-cullet mixture, about 50 % of pollutants in the flue gases like HF, HCl, SOx is absorbed due to chemical reactions with the (earth-)alkali batch compounds [11].

The effect of combustion chamber dimensions and port design on NOx emissions, evaporation (resulting in aggressive vapours towards refractory and particulate formation in the flue gases) and specific energy consumption have been estimated by modelling the combustion process including the gas flows and heat transfer to the melt using own software [13]. Below, results from modelling are summarized for an example case.
Important parameters in this respect are total combustion chamber height, height of burner bank above glass level and volume of combustion chamber.

Example: Design and process optimization of an end-port container glass furnace

CFD modeling has been used to investigate designs of an end-port fired regenerative glass furnace to obtain conditions with high heat transfer and low NOx emissions. During the design-optimization study, boundary conditions like required glass melt quality, high furnace energy efficiency, low CO levels, and low evaporation from the glass melt have been taken into account.

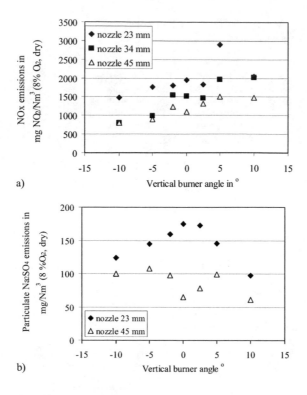

Figure 9   Impact of the vertical burner angle on the NOx formation (graph a) and sodium evaporation rates (b) in an end-port regenerative soda-lime silica glass furnace. The evaporating sodium species (mainly NaOH) will form Na_2SO_4 condensation (fine dust particles) upon cooling of the flue gases.

The CFD model developed by TNO calculates, the kinetics of combustion, temperatures, heat transfer and reactions forming CO and NO or NO_2 in the combustion process.

Apparently, according to the model study (and confirmed by practical observations) for this example of an end-port fired regenerative furnace, maximum reduction of NOx formation can be achieved by burner settings (nozzle size, burner angle), such that long slow mixing flames will occur, in combination with a high crown that will allow recirculation of flue gas above the flame in order to decrease the maximum flame temperature and dilute the oxygen in the root of the flame.

In **figure 9** the calculated NOx and Na_2SO_4 emissions (dust, from evaporation of sodium species from the glass melt surface) are given for a certain example as function of the vertical burner angle. Decreasing the burner angle appears to be beneficial for lowering the NOx formation, but can lead to higher NaOH evaporation rates [18] and thus will increase Na_2SO_4 formation (dust). There seems to be a strong maximum for the high momentum flame of dust emission versus burner angle around $0°$

(horizontal position) for this very case (figure 9b). Due to high dust levels, the regenerator channels might become fouled and clogged. Thus, a low momentum flame is preferred.

The conclusions specific for this end-port fired furnaces according to extended modeling calculations, are:

Nozzle diameter
For limiting the formation of NOx, it is recommended to use a large nozzle diameter, for energy efficiency, beyond a certain size, the nozzle diameter seems to have hardly any effect. A nozzle diameter of around 45 mm in this case resulted in natural gas velocities of about 50 m/s and seems to be the optimum according to the modeling results. According to these results, using burner nozzles with large diameters instead of small diameters also leads to lower sodium evaporation rates.

Burner angle
For this specific case, an almost horizontal burner angle seems to be optimal. Normally a burner angle between 8 and 10 degrees is optimal for energy efficiency, in combination with low-NOx flames.
Fuel distribution
For low fuel gas injection velocity (large burner nozzles), NOx formation and energy efficiency appeared to be not very sensitive to number of burners and the fuel gas distribution among the burners. However, for high gas injection velocities (small burner nozzles), the natural gas of the end-port fired furnace should preferably be distributed over three nozzles. Combustion efficiency does not change significantly for different number of burners, but one single burner causes more sodium evaporation than in the case of 3 burners, due to higher gas velocities close to the glass melt surface in the one-burner situation.

Crown height
Preferably, in this case the ports (and crown) have to be extra lifted by at least 125 – 250 mm, which will lead to a NOx emission reduction of 15 – 20% and only a slight energy efficiency reduction of 0 – 1.5%.

NOx
End-port fired furnace vertical cross section at 25 % from furnace length from port

NOx scaling in mole fraction

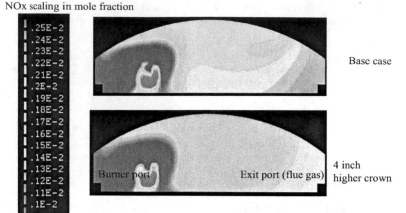

Base case

4 inch
higher crown

Lower NOx-concentration in exit
Reduction 15-25 % in lower case

Figure 10.   Results of simulation of NOx emissions/concentration in end-port fired furnace by
CFD-NOx modeling [13]

Generally, the sodium evaporation rates from the glass melt will also decrease as combustion space volume increases.

Port size
Increasing the burner port size, for instance by an additional 100 mm height of the port leads to a 15% NOx formation decrease, but the combustion efficiency decreases about 1.4 % according the modeling. An increase of the port size is recommended to reduce air velocities and to retard the mixing rate with the natural gas. The evaporation rates will also decrease.

5.   CONCLUSIONS
Energy consumption levels in the container glass industry by optimum furnace design and regenerator size plus controlled operation can be limited to 3.5-3.8 GJ/ton molten glass, depending on the cullet% in the batch. Values below 3.5 GJ/ton are possible when applying batch & cullet preheating. The NOx emissions can be reduced at the same time by changes in port and combustion chamber design, optimization of burner settings and number of burners in combination with tight control of O_2 and CO values in the exhaust gases. Values below 600-650 mg/Nm^3 are feasible, depending on the CO-level permitted in the flue gases (the acceptable CO level depends on the choice of refractory materials in the checkers).
Further major overall energy efficiency improvements are hardly possible without changing the glass furnace concept, and even then the thermodynamic limitation are still valid (range of 2-2.6 GJ/ton molten glass).
Measures such as changes in furnace design, process parameters and raw materials (batch) may impact not only energy consumption and NOx emissions, but may also effect dust emissions, glass quality and

furnace lifetime. Simulation models that take these features into account are indispensable for future developments without taking major risks by design mistakes..

6. FUTURE DEVELOPMENTS AND NEXT STEPS:

In a future project we are proving that an air-fired container glass furnace, based on this combination of technologies will show energy consumption levels between 2.8 and 3.5 GJ/ton molten container glass (dependent on cullet % in batch) with NOx emissions below 0.8 kg/ton molten glass or below 600-650 mg/Nm^3. Seed counts less than 10 per 100 grams are probably achievable with these furnaces and furnace performance. Evaporation of alkali vapours can be limited by lowering the momentum of the flames and increasing the distance between the flames (burners) and glass melt surface. Lower alkali vapour pressures in the combustion space will decrease the attack of superstructure and regenerator refractory materials and lower furnace ageing effects.

7. LITERATURE REFERENCES

1. R. Beerkens: *New Concepts for Energy Efficient and Emission Friendly Melting of Glass.* Advances in Fusion and Processing of Glass, July 11-14, Cairns, Australia
2. W. Trier: *Glasschmelzöfen, Konstruktion und Betriebsverhalten.* Springer Verlag Berlin, Heidelberg, Tokyo, New York, 1984, ISBN 0-387-12494-2.
3. U. Kim; W.M. Carty; C.W. Simon: *Selective Batching for Improved Commercial Glass Melting.* Advances in Fusion and Processing of Glass III, July 27-31, Rochester NY, Ceram. Transactions Vol. 141 Am. Ceram Soc. Pp. 99-106
4. L. Němec; P. Cincibusová: *Glass melting and its innovation potentials: the role of glass flow in the bubble-removal process. Ceramics – Silikáty* (2008) **52**, no. 4. pp. 240-249
5. J. L. Barton: *Innovation in glass melting.* Glass Technol. (1993) **34**, no. 5 pp. 170-177
6. L. Huisman; J. Reijers; R. Brugman: *Advanced and Supervisory Process Control of Glass Furnaces.* XXIV ATIV International Conference – 59th NGF Annual Meeting: Today's Challenges for Glass, July 9-10, Parma, Italy
7. A. Lenhart; H.A. Schaeffer: *Elektrochemische Messung der Sauerstoffaktivität in Glasschmelzen.* Glastech. Ber. **58** (1985) nr. 6, pp. 139-147
8. P.R. Laimböck; R.G.C. Beerkens: *Oxygen Sensor for Float Production Lines.* The Glass Researcher in Am. Ceram. Soc. Bull. **85**, no. 5 (2006) pp. 33-36
9. D.D. Dang; O. Bjorøy; P. Kaspersen: *Measurement of CO and O2 in Gases at High Temperatures.* Proceedings of GlassTrend seminar: Advanced Sensors & Control in High Temperature Processes, 4-6 October 2010, Maastricht, The Netherlands
10. A.M. Lankhorst; A. Habraken; M. Rongen; P. Simons; R.G.C. Beerkens: *Modeling the quality of glass melting processes.* Paper presented at 70th Conference on Glass Problems 13.-14. October 2009, Columbus Ohio, USA
11. R. Beerkens: *Energy Saving Options for Glass Furnaces & Recovery of Heat from their Flue Gases - And Experiences with Batch & Cullet Pre-heaters Applied in the Glass Industry.* 69th Conference on Glass Problems, Columbus Ohio, USA 3. & 4. November 2008, pp. 143-162 Ed. Charles Drummond III, Am. Ceram. Soc. (2009) Published by John Wiley & Sons, Inc. Publication
12. M. Lindig: *Das Integrierte Konzept zur Gemengebehandlung am Ofen.* 85. Glastech. Tagung, Saarbrücken, 30. Mai-1. Juni 2011, Kurzfassungen. pp. 41-42
13. A. Lankhorst, H. van Limpt, A. Habraken, R. Beerkens: *Simulation study of impact furnace design on specific energy consumption, NOx emission levels, volatilization rates and refractory corrosion.* 10th ESG Conference and 84th Annual DGG meeting, Glass Trend seminar on Glass Furnaces and Refractory Materials, 30. May – 2. June 2010, Magdeburg Germany

14. R.G.C. Beerkens; J.A.C. Van Limpt; G. Jacobs: *Energy efficiency benchmarking of glass furnaces.* Glass Sci. Technol. **77** (2004) no. 2, pp. 47-5
15. J.A.C. Van Limpt, R.G.C. Beerkens: *Energy efficiency in glass production.* HVG Fortbildungskurs: Energieverbrauch und Energierückgewinnung in der Glasindustrie, 145-208, Offenbach, ISBN: 978-3-921089-59-0 (2010)
16. R. Conradt: *Zusammenhang zwischen dem theoretischem Wärmebedarf der Reaktion Gemenge-Schmelze und dem minimalen Wärmebedarf eines Schmelzaggregates.* 77. Annual Meeting of the German Society on Glass (DGG) 26.-28. May 2003 Leipzig, Kurzfassungen, pp. 13-18
17. R. Stormont: *Electric melting technologies for energy efficiency and environmental protection.* Glass Worldwide (2008) july/august, no.18. pp. 29-31
18. R. Beerkens, J. van Limpt: *Impact of glass furnace operation on evaporation from glass melts.* Ceramic Engineering and Science Proceedings **22** part 1, 175-204 (2001)
19. R. Ehrig; J. Wiegand; E. Neubauer: *Five years of operational experience with the SORG LONOx melter.* Glass Sci. Technol. (Glastech. Ber.) (1995) **68**, no. 2, pp. 73-78
20. L. Němec; P. Cincibusová: *Glass melting and its innovation potentials: the role of glass flow in the bubble-removal process.* Ceramics-Silikáty (2008) **52**, no. 4 pp. 240-249
21. J.F. Simon; O. Douxchamps; J. Behen ; Y. Joumani : First oxygen fired float glass furnace equipped and operated with a new heat recovery technology. 10th ESG Conference, 30. May-2. June 2010, Magdeburg
22. D. Rue; W. Kunc; G. Aronchik: *Operation of a Pilot-Scale Submerged Combustion Melter.* Proceedings 68th Conference on Glass Problems. Ed. Charles Drummond, III. 16.-17. October 2007, Columbus OH. pp. 125-135
23. R. Gonterman; M. Weinstein: High-Intensity Plasma Melting. Personal Communication Plasmelt Glass Technologies LLC, Boulder Colorado
24. S. Inoue: *Future of glass melting through the in-flight melting technique.* Advances in Fusion and Processing of Glass, July 11-14, Cairns, Australia
25. S. Wiltzsch; H. Hessenkemper: *Erste experimentelle und numerische Simulationsergebnisse zum Segmented-Dünnschichtschmelzer.* 85. Glastechnische Tagung Deutsche Glastechnische Gesellschaft, Saarbrücken 30 May-1. June 2011 Kurzfassungen pp.55.
26. V. Tonarova; L. Němec; M. Jebavá: *Bubble removal from glass melts in a rotating cylinder.* Eur. J. Glass Sci. Technol. A (2010) 51, no. 4, pp. 165-171
27. D.J. Bender; J.G. Hnat; A.F. Litka; L.W. Donaldson Jr.; G.L. Ridderbusch; D.J. Tessari; J.R. Sacks: *Advanced glass melter research continues to make progress.* Glass Industry (1991), March, pp. 10-37
28. R. Kitamura; H. Itoh; Y. Takei; T. Kawaguchi: *Mathematical model of the bubble growth at reduced pressures.* Proc. XIX Int. Congr. Glass Vol. 2. Extended Abstracts, Edinburgh, Scotland, 1-6 July 2001, pp. 361-362

MATHEMATICAL MODELING TO OPTIMIZE A FURNACE LENGTH BY WIDTH RATIO

Erik Muijsenberg[1], Marketa Muijsenberg[1], Tomas Krobot[1], and Glenn Neff[2]
[1] Glass Service, Inc., Vsetin, Czech Republic, [2] Glass Service USA, Inc., Stuart, Florida, USA

ABSTRACT

Mathematical Computer Modeling can be utilized to determine the optimum furnace length by width ratio. Several important furnace issues are related to the furnace length by width ratio, including the type of glass, the type of fuel being utilized, whether the fuel is natural gas or fuel oil, and the maximum pull rate upon the furnace. Each of these factors can impact the glass quality that can be obtained. The mathematical model will contrast these parameters to illustrate the impact to the glass quality based upon the furnace design and its operation.

INTRODUCTION

Mathematical Computer Modeling can be utilized to determine the optimum furnace length by width ratio. Glass Furnace Mathematical Model Simulations have been available to the glass industry for twenty years now, with most all furnaces utilizing modeling for new or redesigned furnaces in today's world.

Modeling is now used by the major glass producers and even the major suppliers to the glass industry to support decisions in glass furnace design changes, because once a furnace is constructed today, one must live with that design for the furnace campaign. For example, a glass container furnace may have a life time of 10 to 12 years, while a float furnace might be 14 to 18 years. Hence, in a career, one might have two (2) chances, and usually no more than three (3) chances to design a furnace before his/her retirement. So, it's appropriate to do some modeling before a new furnace is designed and built.

The factors which should considered before designing and constructing a new furnace are as follows:

- In glass melting processes, significant energy savings can be achieved by relatively small optimizations of the furnace design and/or operating parameters.
- Numerical Computational Fluid Dynamic (CFD) modeling tools can be easily used to investigate changes upon the furnace design and its impact upon energy consumption and glass quality.
- This study will demonstrate the furnace dimension factors that affect the glass melting quality and energy efficiency.

For a given furnace efficiency optimization, the pull rate upon the furnace and its furnace geometry must be considered. In other words, how much more energy efficient can the furnace be made?

Many factors impact a furnace operation, such as the following:

- Specific pull rate and maximum pull rate

- Raw materials / cullet
- Furnace design, total surface area, barrier, deeper refiner, bubbling, etc.
- Combustion system, flame emissivity, burners, stoichiometry
- Regenerator size
- Partial electric heating, or boosting
- Insulation and sealing materials
- Raw material preheating
- Intelligent control (advanced control)
- Often forgotten, is furnace **Length/Width Ratio & Depth**

Glass producers often forget to look at the aspect ratio or the optimum length/width ratio. Of course, certain factory physical limitations will limit the changes that can be made to a furnace.

We will focus upon the length/width issues for a certain furnace type and consider a container furnace for our review, since most furnaces are such. The general concept would still apply to other furnaces, even if the furnace is a float furnace, or fiber furnace. In general, the concept would be the same for other furnace aspect ratios.

One of the items to optimize upon the furnace would be energy efficiency, which includes the combustion system, the flame emissions, and the stoichiometry of burners. We can also look to partial energy improvements such as electric boosting, or preheating of the raw materials, regenerator sizing, or advanced furnace controls, but these items will be addressed separately.

We can learn a lot from the nature. At least we can achieve a basic understanding about why we do things in a certain way. An alligator is long and narrow, see Figure 2. A penguin, by contrast, is short and wide. See Figure 1. The outer surface of the alligator is about ten times larger than the penguin, and the temperature gradient is approximately 0°C. On the other hand, the penguin has a temperature gradient of about 100°C. So, different animals have different length to width ratios, and the respective volume to surface issues.

Figure 1 [1]   Figure 2 [2]

Let us now look to several dimensional cases. If we have three (3) cubical cases, such as a dimension of 1 x 1, 2 x 2, and 3 x 3, one can readily see a difference of a factor of three (3). There is an increased surface area with the smaller cube, which means a potentially high heat loss.

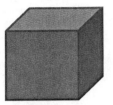

Surface 6 m^2
Volume 1 m^3
S/V   6 m^{-1}

Surface 24 m^2
Volume 8 m^3
S/V   3 m^{-1}

Surface 54 m^2
Volume 27 m^3
S/V   2 m^{-1}

Figure 3

These are cubes; so let us look to rectangular dimensions as well. If the middle cube shape is made two (2) times longer, and one (1) times less wide, with an aspect ratio quite close to 1, we would have a surface to volume ratio of 3.5. If we make it even longer, we will have even more surface area, and not so much volume, resulting in even more heat losses.

Hence, the rectangular shapes show a lesser improvement than the cube shapes, and the surface to volume ratio becomes greater and less desirable.

Surface 28 m^2
Volume 8 m^3
S/V   3.5 m^{-1}

Surface 42 m^2
Volume 8 m^3
S/V   5.25 m^{-1}

Figure 4

The conclusion would be that larger is better, and hence the larger the furnace, the more efficient it will be. The more one stays like the penguin than the alligator, the more energy efficient one can be.

Another extremely powerful comparison from nature, are the Figures 5 and 6 that show the volumes of the earth and the sun, respectively. The earth over its billions of years in existence is comparably hot inside, and due to the smaller volume and small surface, the surface to volume ratio is 10^{-7}. The sun is even more efficient with the surface to volume ratio of the earth to the sun of 10^{-9}. Of course, we would not build such a big glass furnace, but if we could, we should look for this kind of ratio. In other words, the smaller the surface/volume, the more heat is retained inside.

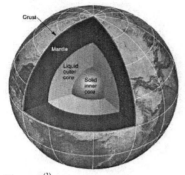

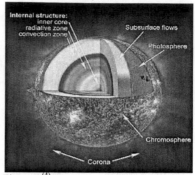

Figure 5 [3]
Earth
Surface 5×10^{14} m^2
Volume 1×10^{21} m^3
S/V 5×10^{-7}

Figure 6 [4]
Sun
Surface 6×10^{18} m^2
Volume 1×10^{27} m^3
S/V 6×10^{-9}

Shown below are two (2) examples of contrasting model studies in Figures 7 and 8.

On the one hand, we have a typical float furnace of approximately 1,386 square meters, (5,920 square feet) of surface area, 900 cubic meters (31,783 cubic feet) of volume, and a surface to volume ratio of 1.54. This furnace is very large.

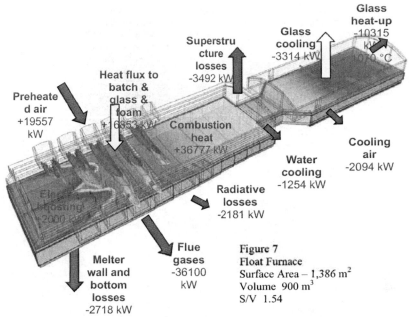

Figure 7
Float Furnace
Surface Area – 1,386 m^2
Volume 900 m^3
S/V 1.54

On the other hand, an electric melter, which is quite a bit smaller than the prior example, has a good surface to volume ratio. In this comparison, the electric melter is only 110 square meters (1,184 square feet) of surface area, and about 75 cubic meters (2,649 cubic feet) in volume, which is an even better surface to area ratio of 1.46. Electric melters are relatively energy efficient.

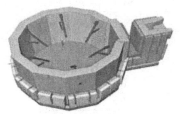

Figure 8   Electric Furnace
Surface Area - 110 m^2
Volume 75 m^3
S/V 1.46

Let us review typical container furnaces, to compare their respective energy efficiencies. We can see the typical furnaces, including cross-fired furnaces, end-fired furnaces, oxy-fuel furnaces, and small furnaces. As these are European container glass furnaces, the efficiencies include 60% cullet. As we consider furnaces with larger volumes, the furnace efficiency improves.

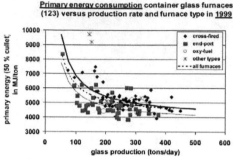

Figure 9 [5]

When looking at the furnace types above, one can see that the end-port furnace has the better energy efficiency. The regenerator surface to volume ratio for the end-fired furnace is much better than that of the cross-fired furnace. Also, the end-fired furnace typically has an aspect ratio that is much shorter and wider than for the cross-fired furnace.

As shown in Figure 10, it should be noted that a furnace with a higher pull rate also has a better efficiency.

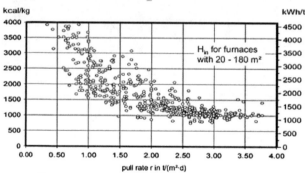

Figure 10 [6]

For our comparisons, let us review a typical end-fired furnace or a "U" flame furnace. This comparison furnace will be under port fired using four (4) burners, but of course three (3) or two (2) burners could be used, depending upon the furnace design and the emissions that are allowed. Our furnace will operate with 20% cullet, with a production rate of 240 metric tonnes per day. The furnace is expected to operate utilizing about 1,380 cubic meters of gas per hour of natural gas, or an equivalent furnace efficiency of about 5.0 mega-joules per kilogram of glass. In our example, the amount of cullet will be closer to US standards at about 20%.

Shown is a standard design for an end-fired furnace:

- End-fired furnace (U flame)
- Four (4) burners
- Clear glass, 20% cullet
- 240 MTPD
- 1,380 Nm³/hr. natural gas (5 MJ/kg)
- Length/width ratio 1.54
- Surface/volume ratio 1.74
- Note: All model cases with the different L/W versions, are utilizing the same relative batch lengths

Figure 11 shows a typical standard furnace length to width ratio of 1.54:1, and a surface area to volume ratio of 1.74:1, including the views of the flames, the batch pattern, and cross sectional views of the glass melt.

Base Case – Flames   Base Case – Batch Pattern

Base Case– Glass Melt

Figure 11

For the Base Case that is a typical end-fired furnace with under port burners, the gas is mixing with the combustion air for combustion. It is important to keep the combustion inside the combustion chamber without flame impingement on the end wall. It can be seen that the Base Case shows a well-placed length of the flame.

The batch is coming in from the two (2) doghouses with the batch moving into the center of the furnace. There is also a barrier, near the throat of the furnace, with the batch moving according to the convection patterns, and ultimately over the barrier and into the deep refining area. The glass then goes into the throat, and then to the forming machines.

Of course, there are many furnace design options, but in reality one must consider the existing furnace hall and the physical limitations thereto. A deeper furnace can also be considered, but generally the position of the forming machines may cast some limitations. Additionally, it is important when we change the length to width ratio that the batch remains in the relative center of the furnace.

Figure 12 illustrates the Base Case with an aspect ratio of 1.54:1 and a width of 8.0 meters, which we will compare with a wider furnace at 9.0 meters and an aspect ratio of 1.21:1, and then to a Super Wide Model of 10.0 meters and an aspect ratio of 0.98:1.

When considering a float furnace it could be more or less a ratio of about 3:1, with the average float furnace (melter). We are now seeing wider float furnaces in Europe, which are very clearly becoming more energy efficient. If you look at the surface area to volume ratio, at a certain point you can see that it doesn't change so much, but the surface area to volume ratio is smaller for that of a very large furnace, and is also good for the glass quality.

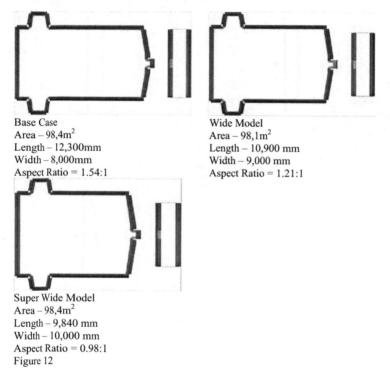

Base Case
Area – 98,4m^2
Length – 12,300mm
Width – 8,000mm
Aspect Ratio = 1.54:1

Wide Model
Area – 98,1m^2
Length – 10,900 mm
Width – 9,000 mm
Aspect Ratio = 1.21:1

Super Wide Model
Area – 98,4m^2
Length – 9,840 mm
Width – 10,000 mm
Aspect Ratio = 0.98:1
Figure 12

Now we will make the furnace longer as shown in Figure 13, with a length to width ratio of 2.0:1, which is a quite common situation for cross fired furnaces, and then to a Super Long Model of 16.4 meters and an aspect ratio of 2.73:1.

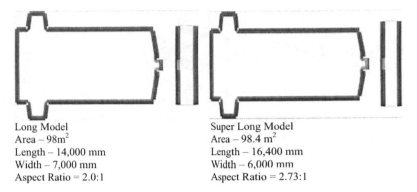

Long Model
Area – 98m^2
Length – 14,000 mm
Width – 7,000 mm
Aspect Ratio = 2.0:1

Super Long Model
Area – 98.4 m^2
Length – 16,400 mm
Width – 6,000 mm
Aspect Ratio = 2.73:1

Figure 13

So, now the question is which furnace aspect ratio is better? Then, "how much better" is the furnace? To answer this question, let us use modeling to determine how much energy can be saved, and what the glass quality will be. Modeling can determine this for us.

From Figure 14 below, and the side view of the base case, it can be seen that the wider model starts to look short from the side view. The simple wider model starts to look even shorter. And when viewing the longer model and the super long model in Figurer 15, one can really see that the super long model starts to look like a float furnace. It looks narrow and long.

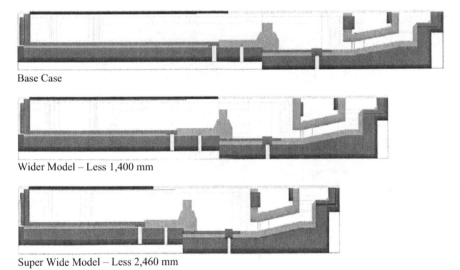

Base Case

Wider Model – Less 1,400 mm

Super Wide Model – Less 2,460 mm

Figure 14

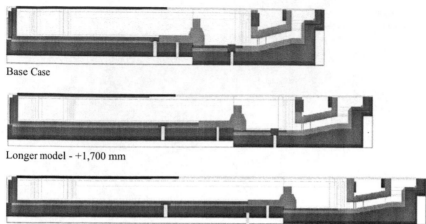

Base Case

Longer model - +1,700 mm

Super Long Model - +4,100 mm

Figure 15

Modeling can demonstrate what is happening, but to have a valid comparison we need to keep certain parameters the same, in particular the fining of the glass and hence glass quality. Therefore, we want to keep the glass temperatures the same with a change to the amount of gas. However, we will keep the refining temperatures the same, and thus the same for the refining area as well.

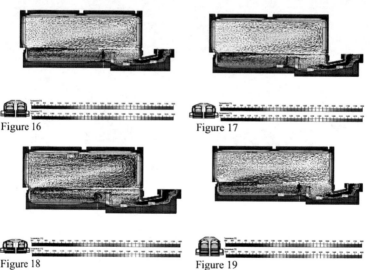

Figure 16   Figure 17

Figure 18   Figure 19

Figure 20

Although not precisely the same temperatures, the glass temperatures in the throat are 1,348°C in the Base Case (Figure 16), 1,349°C in the Wide Model (Figure 17), 1,354°C in the Super Wide Model (Figure 18), 1,347°C in the Long Model (Figure 19), and 1,349°C in the Super Long Model (Figure 20); hence, the throat temperatures are essentially the same.

Figure 21 shows some pictures from the different cases, including the Base Case, the Wide Model, and the Super Wide Model. In general, the illustrations do not, of course, change much between the wider and shorter furnaces and the Base Case, but there is more flame contact upon the bridgewall of the furnace. This flame impingement is at the limit for a real furnace operation. Using the modeling and comparing the furnace efficiencies, one can clearly see the differences. We also have to be sure of the batch flow pattern, with more flow going towards the center of the furnace.

Base Case

Wide Model

Super Wide Model; Risk of overheating opposite wall
Figure 21

So if we go the other way, and refer to the top views of the furnaces of the longer model and the super longer model, it can be seen that there is flame impingement upon the side wall. Also, of course, it takes a little bit more time for the batch to get to the back to the center of the furnace.

Base Case

Longer Model

Super Long Model
Side wall risk for overheating
Figure 22

Shown in Figure 23 are other top views but now showing the iso-lines for the temperatures. The top view of the Base Case shows a nice flame distribution, with the hot spot before the end wall, and with the flames fairly close to the end wall. In the super wide model it really starts to be close to the end wall, and could be risky. To avoid a real problem in actual operation, the burners would have to be tuned. On the other hand, there is no overheating on the side walls.

Base Case
Doghouse Temperature: 1,211°C

Wide Model
Doghouse Temperature: 1,176°C

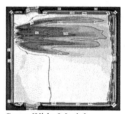

Super Wide Model
Doghouse Temperature: 1,159°C
Figure 23

In Figure 24, for the super long melter we should perhaps use three (3) instead of four (4) burners. Already, there could be overheating upon the side wall, but no overheating for the end wall. And in the case of the super long melter, it can be said that the flames no longer cover even half of the melt surface. These effects produce a lower fining index, and hence is an indicator of lower glass quality.

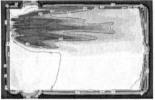

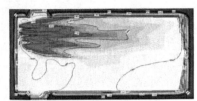

Base Case   Longer Model

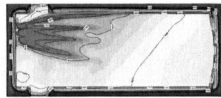

Super Long Model
Figure 24

The best way to understand what these differences mean for the glass melting performance is by tracing mass-less particles over the melting process. We follow the temperature and viscosity histories, and they give us the (minimum) residence time, and the melting, mixing, and fining indices. These indicators are a good indication of what is happening to the glass quality.

We will also trace the bubbles. So, when we release bubbles from the batch blanket, (Figures 25 & 26), we will see how many of them reach the throat and become a potential defect.

In Figures 27 and 28 we can see, of course, many of these bubbles are refined. Many of these bubbles are refined in the melting zone, with some finally fined in the refiner. Figure 29 shows the bubbles that did not refine, which come into the glass product.

Starting position of the bubbles-50/m2 under the batch
CO_2 bubbles (about 20 000) of a size 0.1 – 0.5 mm were placed under the batch area and were calculated. Time of calculation was 48 hours.

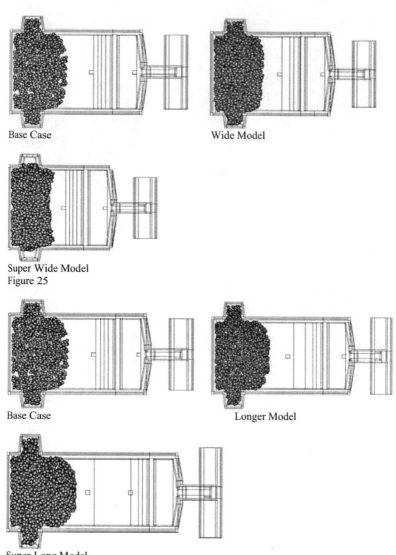

Base Case

Wide Model

Super Wide Model
Figure 25

Base Case

Longer Model

Super Long Model
Figure 26

In the case for the longer melter, one can see that the batch blanket gets longer. And when we want to see where the bubbles are refining in the longer melter, we can see more distribution but not so many bubbles.

Conditions for the Glass Model Indices

- Glass quality parameters are derived from particle tracing prepared for the model of the melting part.
- The evaluation of the melting performance can be done on the basis of tracing and the mass balance computation. About 50,000 tracers as mass-less particles were inserted to the batch chargers and were traced within a furnace. Trajectories of those tracers and time when they passed the goal in the canal end were recorded.
- For each particle the *residence time* in the glass model is calculated. The fastest particle with the minimum residence time is called the 'critical trajectory'.
- Besides the time, we also calculate the temperature, viscosity and velocity along each trajectory. From this information, the melting index, fining index and mixing index are calculated for each trajectory.
- The *melting index* indicates the quality of the melting process along the trajectory: a high value means that the particle resides for a long time in regions with high temperature and low viscosity.
- The *fining index* indicates the quality of the fining along the trajectory: a high value means that the particle resides for a long time in regions with a temperature above the typical fining temperature and low viscosity.
- The *mixing index* indicates the quality of mixing or inhomogeneity dilution along the trajectory: the value denotes how many times a 1 cm thick cord can be diluted along the trajectory.

Starting position of the bubbles-50/m^2 under batch
CO_2 bubbles (about 20,000) of size $0.1 - 0.5$ mm were placed under the batch area and were calculated. Time of calculation was 48 hours. Final position of the bubbles:

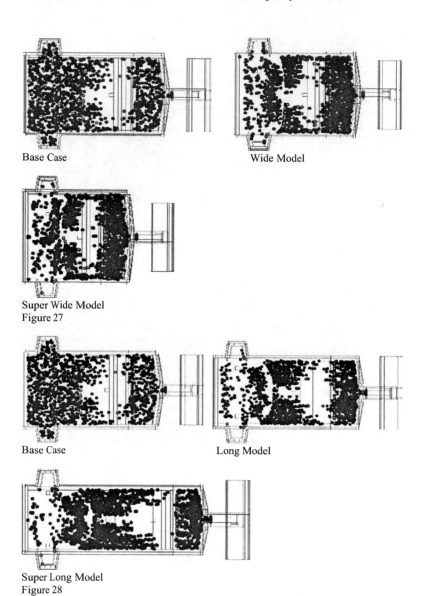

Base Case

Wide Model

Super Wide Model
Figure 27

Base Case

Long Model

Super Long Model
Figure 28

We can see in the Base Case the simple fact that about 2% of the bubbles reached the throat. If the furnace is made a little bit wider, the defects are reduced to about 1%, (more or less the quality is doubled). It becomes 0.3% and better if the furnace is made even wider, showing even more improvement.

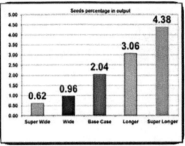

Figure 29

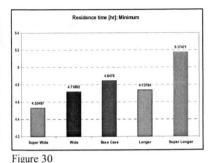

Figure 30

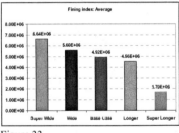

Figure 31

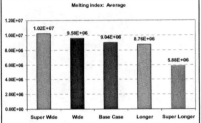

Figure 32

Figure 33

Also the fining index is better for wider melters, with the better flame coverage (especially downstream). In fact the better glass quality can also be used to reduce the energy input.

To maintain the glass temperatures, we had the model automatically adapt the amount of gas that was utilized, as shown in Figure 34. Below we can see that the super wide model uses about 2% less energy and continues to deliver better quality. The super long melter uses almost 4% more energy (to maintain the same throat temperature), but the glass quality would be lower.

	Super Wide Model	Wide Model	Base case	Longer Model	Super Long Model
Gas Amount [%]	-2.11	-1.3	0	+2.91	+3.8

Figure 34

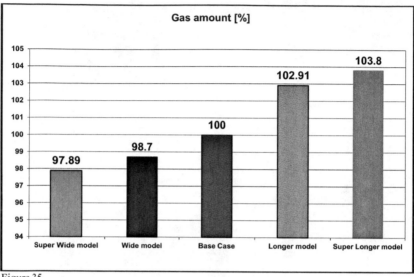

Figure 35

So we may summarize as follows:

- A shorter and wider furnace (close to rectangular) shows the best fining index, and lower seed levels, thus a higher melting performance and better fuel efficiency. This is mostly due to the lower wall surface heat losses and better flame coverage.
- Overheating of the end wall refractories have to be avoided as shown by the wider models.
- The ideal furnace shape depends upon the production hall layout.
- The theoretical optimal 100 m^2 end-fired furnace seems to be sized 10 x 10 m with glass depth around 1,5 m (see the former paper presented at the Glass Problems Conference in 2008). [7]
- Potential fuel savings costs comparing the longer furnace versus the wider furnace are: 6% energy savings which, for this example, equates to €220,000 per year and over the furnace campaign of 10 years: €2,200,000.
- The return on Investment (ROI) of such a Model Study is extremely quick!

We may conclude that for the end-fired furnace the optimal length to width ratio is probably around 1.2:1 whereas most furnaces are in the range of 1.8:1.

Last and not least, we have to remind ourselves that each furnace case is individual and the furnace studies shown here are just a demonstration how an issue such as the aspect ratio can lead to significant effects upon the melter performance and its energy efficiency.

References:

1. Emperor Penguin, True Wild Life online, 22 February 2011
2. Wigmore, B., Daily Mail online, 13 June 2006, "Darling there's an alligator at the door"
3. Courtesy of Lawrence Livermore National Laboratory, University of California, Livermore, CA, Penrose C. Albright (Director)
4. Follow, T., "Getting to Know Your Solar System": The Sun, updated 12 August 2011, Rendering of the solar core via University of Northern Colorado
5. Beerkens, R., "The Most Energy Efficient Glass Furnace: Energy Efficiency Benchmarking and Energy Saving Measures for Industrial Glass Furnaces", presented at the 21st ATIV Conference in Parma, Italy, 21-22 September 2006
6. M. Lindig, "The Challenge of Conventional Furnace Design," 7th ESG conference on Glass Science and Technology, Athens 25-28th April 2004
7. H.P.H. Muijsenberg, J. Ullrich, G. Neff, "Laboratory Experiments and Mathematical Modeling Can Solve Furnace Operational Problems", presented at the 69th Conference on Glass Problems, The Ohio State University, Columbus, OH, November 4-5, 2008

A SUMMARY OF ALMOST 50 YEARS OF GLASS FURNACE PREHEATING

George Kopser
Hotwork-USA

ABSTRACT

This paper will cover the almost 50 years of preheating glass melting furnaces with the Hotwork technology that was invented in 1962 by Mr. Trevor Ward of Dewsbury England, UK, and to which we refer to as pressurized hot air heating. We were awarded the license to use this technology in North and South America in October 1965. I joined the company in January 1966 and was trained by Mr.Ward and others as how to build and operate the equipment.

The technology is centered around a unique burner that Mr.Ward developed that could fire with a flame as tiny as that on a home cook top range but with tens-of-thousands of cubic feet of air pressure passing by it. This provided the opportunity to pressurize the furnace by using air as the medium. This enhances temperature uniformity while at the same time allowing safety devices like automatic ignition, flame failure detection and fuel shut off devices which are difficult to apply to torches or piles of wood and coal that were used previously.

PIPE BURNER HEATING

Figure 1: This is a photo of a gas pipe or stick burner as they were called that was provided by Mr. Phil Ross on one of the last glass furnace heatups that he did before the introduction of our technology. Many of these pipe burners were used on even small furnaces in an attempt to eliminate hot spots near and above the burner flame.

Mr. Ron Walton told me that on his last large plate glass furnace heatup , and I quote "We used 80 stick burners , two 1" black iron pipes stuck through the tuck space from each side of the furnace, all 4 feet apart the entire 160 foot length of melter and working end. We pulled them back 4" per day for the 30 days of heatup, except the ones in the port area which we removed when we went on port gas" end of quote.

And, even with this much effort, there was no guarantee that the refractory was not damaged on the newly rebuilt furnace!

Explosions also occurred due to the lack of ultra violet monitoring of the flames and sometimes crowns would collapse due to the severe temperature non-uniformity from side to side.

TYPICAL BURNER SETUP

Figure 2: This is a photo of a standard set of heatup equipment and the number of burners used is dependent on the size and complexity of the furnace. This shows the burner, combustion air fan, and fuel train with all of the safety gear inside.

Because the torches, or piles of burning wood or coal did not have a combustion air supply they relied on a negative pressure in the furnace so that cold air could be drawn in, which increased the internal temperature non-uniformity.

By using the pressurized hot air method, the furnaces were heated more uniformly and furnace downtime was significantly reduced. This method uses what some refer to as excess air high velocity burners to provide a forced convection type of heating. We don't like to use the word "excess" as it denotes "wasteful' or "too much". These burners do use more air and therefore more oxygen than is required for the initial combustion process, but the additional air is used as the medium for the forced convection process. This provides the scrubbing action of dry, warm air across the surface of the refractories.

IMPORTANCE OF A POSITIVE INTERNAL FURNACE PRESSURE

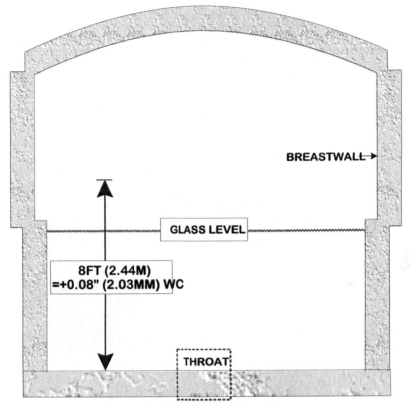

Figure 3: Furnace Pressure – Rule of Thumb.

Internal furnace pressure is one of the most crucial items in obtaining a thorough heatup of the refractories. The furnace pressure must be high enough to drive the heat to the bottom, through the throat and out the forehearths. The pressure must also be constant.

Our rule of thumb is to have +0.01 inch of water column for every foot of height from the throat to the furnace pressure tap.

If the tap is 8 feet above the throat, then the furnace pressure should be at +0.08 inch W.C or higher to ensure that the heat is driven to the bottom.

If pressures below this guideline are attained, it should be noted that most of the heat source for the furnace bottom and throat will be via radiation from the crown. In certain instances this may not be acceptable. If tamped or rammed refractories are present then steam explosions can occur when the heat source is via radiation vs. convective heat. Convection provides more uniform temperatures across the entire refractory surface and starts the drying process much earlier than radiant heat. As the furnaces have grown in size over the years, the pressure taps may have been raised which negates targets used in the past.

When we started in 1965 the owner of Hotwork was my cousin Bob Burger who had worked for a glass bottle manufacturer after graduating from ceramic engineering college. He then went to work for a large refractory manufacturer, so he knew the importance of a proper heatup of the refractory after the furnace rebuild.

Downtime was always a problem for manufacturers and with the help of furnace engineers and large glass manufacturers, Bob developed some very fast heatup curves which we still use to this day. These faster curves would never have been successful using previous heatup methods.

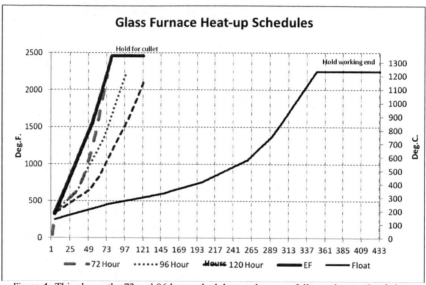

Figure 4: This shows the 72 and 96 hour schedules used successfully on thousands of glass furnaces. I should note that with these fast schedules it is imperative that the personnel doing the rod work and the tools, wrenches, etc. be excellent.

As the bottle, fiberglass, tableware, and TV furnaces got bigger we slowed the schedules to 120 hours. Electric melting with zero silica brick had different schedules based on AZS and later dense chrome refractories. When fused cast "M" Refractories came on the scene we followed those manufacturer's schedules.

Float glass furnace schedules are typically longer mostly due to the physical size and since the owner wants normally just one person supervising all refractory movement and steel adjustments because there are hundreds of jack bolts, pressure plates and cross and longitudinal rods to attend to.

In 1984 Mr. Ward's company went out of business and many of the licensees started doing their own thing.

After Hotwork-USA purchased the remnant of the original English company and one of their operations in Asia/Pacific, we saw that the balance of the world was approaching heatups entirely differently. The schedules were normally much longer and the approach to heating the front ends was not acceptable to us.

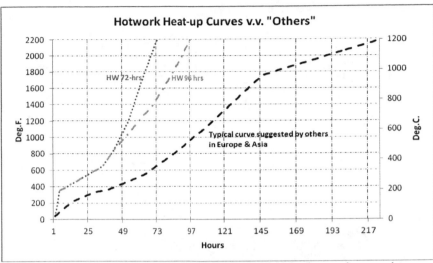

Figure 5: This slide compares the popular 96 hour schedule that we have used to a European schedule that is much longer. The surprising thing is that the designs of the furnaces and the refractories used are very similar and they are sometimes owned by the same global glass manufacturer.

In the early days, the type of fuel sometimes affected the heatup rate. Eventually the English Hotwork improved their oil firing burners to be more stable and capable of following an aggressive schedule. Even after that problem was addressed and even now that natural gas or LPG is available worldwide, I'm still not sure why the schedules are so long. It's not that the furnaces will run longer because of a slower heatup as most of our clients are getting 12 to 15 years of operation under normal circumstances. It's normally the type of glass made or operating problems that shuts them down sooner.

Actually I believe it's cultural in that Americans seem to be more aggressive by nature.

We sometimes find even longer schedules in Asia even though the same heatup equipment and personnel are being used. It's certainly something that needs to be addressed by the industry as even the schedules submitted by the major refractory suppliers are slow compared to what we have been doing for 46 years this month!

EVOLUTION OF HEATUP BURNER LOCATIONS

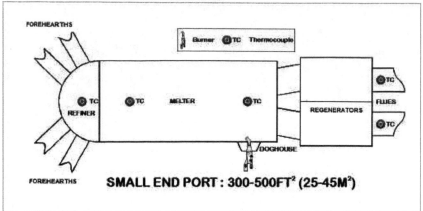

Figure 6: This slide shows what was typical in the USA in the 1960's - either a small side port or end port and normally only one opening large enough for our burner.

There was no client combustion air fan and the combustion air was drawn in by a natural draft stack. It was difficult to obtain internal pressures above +0.05" WC. But since the furnace was so small, the heatup was still better than the stick burner method.

Unit Melter With Recuperator

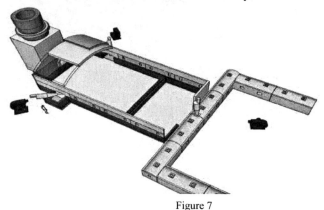

Figure 7

As the front ends become larger and more complicated the clients needed a heat source for those. Here you can see a "T" shaped nozzle in a distributor near the throat. These require a large opening for maximum pressure and temperature uniformity. We also started to use more burners in the melter and by the early 1970's most furnaces had combustion air fans that we used to supplement the air we were pumping in.

End Fired Regenerative Furnace

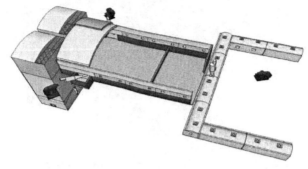

Figure 8: This shows a similar setup on an end port.

The doghouses were getting larger to accommodate the higher pull rates so we started asking for burner openings in breastwalls, endwalls, and backwalls as the burner in the doghouse created a lower temperature area and we want it to get a good heatup too.

Cross Fired Regenerative Furnace

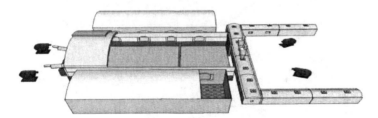

Figure 9: Two 90° outlet nozzles firing left and right.

Some clients started using atmospheric baffles in distributors and forehearths so we had to use two 90° outlet nozzles firing left and right so as to not damage the baffles.

Large TV Furnace w/ Round Refiner

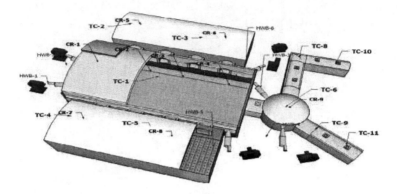

Figure 10

By the 1980's the furnaces were getting much larger but it normally allowed more positions for more heatup burners, which enhanced the temperature uniformity. The checker packs almost doubled to make the furnaces more fuel efficient so we needed more fire power for this.

Float Furnace

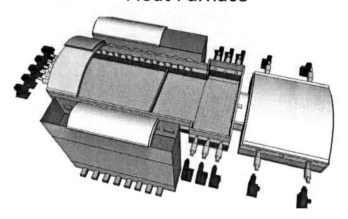

Figure 11

When we first started doing large plate, sheet, and float glass furnaces we used ten burners as that's what experience with smaller furnaces showed us that we needed. Many people said we must use 15 to 25 burners as less burners would shorten the campaign life, but our clients were getting 12 to 17 years furnace life which is about what we later found throughout the World.

We normally use about 15 to 18 burner sets now and it does provide better temperature uniformity but the furnaces don't run any longer. Our clients are now approaching 20+ years, but that is normally due to hot repair maintenance that developed over the years.

Oxy/Gas Furnace w/ "H" Forehearths

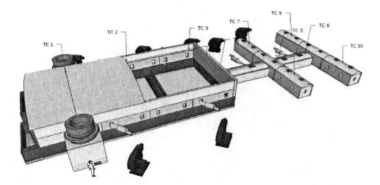

Figure 12: When oxy/gas furnaces came into play, this allowed an even better heatup burner positioning. We recommend a circular rotation for maximum temperature uniformity. You can see the front end burners and both are firing "downstream".

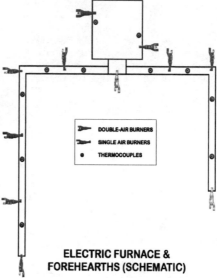

**ELECTRIC FURNACE &
FOREHEARTHS (SCHEMATIC)**
Figure 13

By the 1990's we started to see extremely long and narrow distributors and forehearths. We found that the best rule of thumb is to have a heatup burner every 30' and all waste gas flowing in the same direction either upstream or downstream.

Large Side Port w/ Three Forehearths

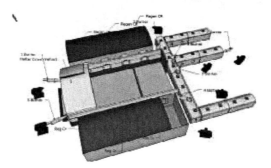

Figure 14: Here is a large system where adding the third forehearth required additional fire power.

Large Oxy Gas Furnace w/Complicated Front End

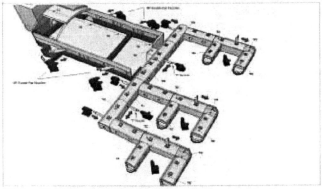

Figure 15: This drawing also shows the circular motion in the melter.

Nowadays the front ends are extremely large and it is important that in the early design stages that properly sized and strategically located openings for heatup burners are planned.

Other considerations must be taken into account as oil cannot be used on the "T" and 90 degree outlet nozzle. These deflector type nozzles can only operate to the 1800°F range without deteriorating. If temperatures above this level are required for all electric front ends or for boosting the temperature at the throat, the "T"'s and 90's must be removed, and a straight nozzle installed.

SILICA CROWN GROWTH

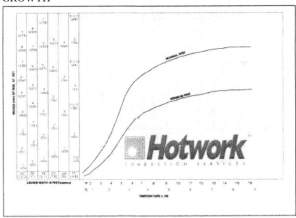

Figure 16: Expected Silica Crown Growth

This last slide is a guideline that we formulated from information gathered over the years showing expected silica crown growth. The vertical axis on the left lists inches of rise at the key line for crowns of various widths, 40 feet wide down to 15 feet wide. The horizontal axis lists the temperature up to 1600° F. The lower curve represents the minimum rise that we would like to obtain on a heatup. The normal rise is shown in the upper curve. For example, if you have a silica crown that is 30 feet wide, then you should expect about 3 inches of rise at 500°F. and about 5.5 inches of rise at the end of the heatup.

CONCLUSION

Since 1962 thousands of glass melting furnaces have been preheated using the pressurized hot air method. This method has allowed manufacturers to build very large and complex furnaces containing a wide variety of refractory material.

The application of preheating equipment has changed over the years, but the original technology has not and has been accepted throughout the world. It is regarded as the state of the art method for preheating glass furnaces.

We are proud of our involvement in helping to introduce this technology to the glass industry and look forward to serving the industry for another fifty years.

IS 50% ENERGY EFFICIENCY IMPROVEMENT POSSIBLE IN THE GLASS PRODUCT AND
PRODUCTION CHAIN IN 2030?

Léon Wijshoff
NL Agency

ABSTRACT

*The challenging question whether 50 % energy efficiency improvement in 2030 is possible, has been
asked by the Dutch Ministry of Economic Affairs, Agriculture and Innovation in 2009 to all industrial
sectors involved in the Dutch Long Term Agreements programme.*

*A short review on the history of the Dutch Long Term Agreement program (LTA) is provided after
which focus is given to the actual status of the LTA and how the Dutch Glass Industry handled the
challenging question above and came with a constructive and focussed answer.*

The start of the LTA programme dates from the late 1980-ties when the Dutch Government was
introducing a stricter energy and environment policy, e.g. for the energy intensive industries. The
government decided that enforcing severe legislation was not the way to reach their targets and choose
for a cooperative approach (e.g. voluntary but challenging agreements) with the industry. Voluntary
agreements were seen as the key-instrument to focus on energy efficiency in the long term.

As industry was open for such cooperation, consultations with the branch associations started on how
to make it operational and how Dutch Government could stimulate innovation and innovative thinking.
By means of the Energy Potential Scan methodology (mainly for production processes); it was possible
to identify the long term potential for energy efficiency improvement in the production process of
individual companies as well as for a whole branch. This potential setting methodology has been the
base for the agreements.

The (first generation) Long Term Agreements between various Ministries and industrial associations,
on behalf of their member companies, were set for the period 1990-2000. The average target in these
LTA's was 20 % energy-efficiency improvement in the production processes. Companies having an
energy consumption over 0.1 PJ were eligible to participate in the LTA's. Participation to LTA is
voluntary but not without obligations. The obligations for the participants are:

- Establish up 4-year energy efficiency plan (EEP), indicating which measures are expected to be
  taken to improve the energy efficiency and their expected energy savings.
- Participate in the yearly energy consumption monitoring programme
- Introduce and maintain an energy management system.

The basic philosophy for LTA is striving for continuous improvement not only in direct energy savings but also in production efficiency and meeting environmental legislations. Government's contribution consisted of financial support in:

- Financial support schemes, e.g. for applied research and demonstration projects of new technologies (e.g. consultancy support for the Energy Potential Scans and Monitoring
- No extra requirements in the environmental permits of the companies
- Financing NL Agency to arrange the monitoring, training and general support to all participants.

In this first LTA period the focus was on energy efficiency in the production processes of the companies. In total approximately 1000 companies were involved in the LTA, representing more than 80 % of the Dutch industrial primary energy consumption

In 2000 the LTA approach showed its success. On average an energy-efficiency improvement (energy savings per unit production) of more than 22 % was achieved compared to 1990. [1]

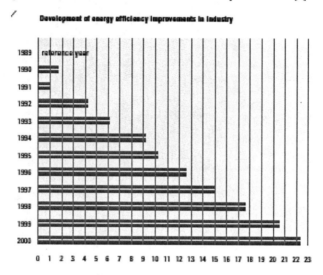

Development of energy efficiency improvements in Industry

A lot of measures were taken, energy efficiency had become a management issue and most of all the fact that LTA linked companies was seen as its success.

The Dutch Glass Industry reached a 16 % energy-efficiency improvement in this period whereas their goal was 20 %. Its total energy consumption increased from 11 PJ in 1989 to 13 PJ in 2000, production increased by about 35% showing that growth was still possible for Industry.

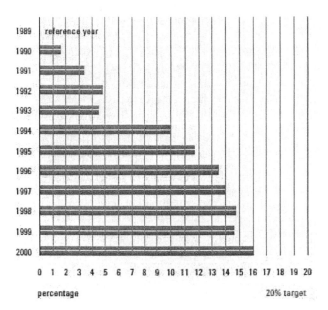

Some of the major achievements in this period were:

- A lot of cooperative research by the Dutch Glass producers carried out with assistance of mainly the TNO-Glass Group in Eindhoven;
- Introduction of the first waste heat recovery unit for batch preheating in container glass;
- Development of a Glass Handbook as tutorial for the Glass industry;
- Further development of oxygen fired glass furnace technology with high energy efficiency and low NOx emissions;
- Process control software development
- Model based furnace control

The positive results of the LTA's and the fact that energy efficiency was still high on the political agenda was a major reason to continue with the cooperative approach. However some changes were made. For the energy intensive industry a specific **Energy Efficiency Benchmarking Covenant** was established

The aim of the Benchmarking Covenant was that the participating companies would belong to the top (10%) of the world regarding energy – efficiency in 2012. Extensive world-wide benchmark studies have been carried out to position the individual Dutch glass factories & furnaces in the ranking of energy efficiency levels.

Companies had to make plans to become among the 10% most energy efficient if this was not already the case.

Government did not provide any financial support to the companies. Their contribution was the cost for benchmarking studies (inventories) [1] and for an independent Verification Body to monitor the progress and agreement. Additionally the Government agreed not to create additional regulation on energy-efficiency or CO_2 emission reduction.

For the medium sized companies a second generation of the LTA programme continued in which the scope was enlarged to include energy efficiency improvement in the whole product chain without quantitative targets.

A recent study of TNO-Glass reported an energy efficiency improvement of 7 % for the Dutch glass industry from 2000 to 2009. [2] The study also showed that nearly 60 % of the energy is required for the melting process. The specific energy consumption is 8,8 GJ/ton molten glass and 9,9 GJ per ton of sold glass.

Changes in the Dutch Governmental policy in 2007 as well as the growing role of the European Union, e.g. the Emission Trading System (EU-ETS) gave rise to establishment of the third generation of Long Term Agreements in 2008.

The LTA-3 distinguishes between EU-ETS and non-ETS companies. For all companies the focus for energy efficiency improvement is on the complete production chain. The target for the **non-ETS** companies is an energy efficiency improvement of 30% in the period 2005-2020, an average of 2% per year, aiming at 20 % at company level and 10 % outside the company, in the product chain. The latter might include energy efficiency improvements in the raw material phase , the product use phase or the recycling phase as well as application of renewable energy sources, see figure.

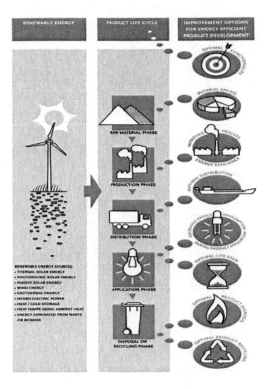

For the **ETS** companies, amongst them all glass companies, no target were set, but the companies are expected to implement all energy saving measures with a pay-back period of less than 5 years.

In this LTA3 covenant, one of the obligations or challenges for the companies and their sector associations is the development of a Roadmap. The Roadmap is meant to provide a strategic vision up to the year 2030 and indicates which technological and non-technological issues need to be tackled to enable a potential 50% energy efficiency in the whole product chain improvement in 2030 compared to 2005. This counts for almost all glass products: flat glass, insulation wool, glass fibres, domestic glass, container glass.

As indicated, the LTA3 programme was launched in 2008. Preparations of Roadmaps started in 2009. Despite the fact that the economic crisis of 2008 forced many companies to focus on survival as production fell deep, many companies used this period as a moment to reconsider their future and identify where new opportunities could be.

The main factories of the Dutch Glass industry participate in the ETS LTA-3 Agreement, covering 6 companies with 9 production facilities. The product range covers container glass (80 % of the mass of the production), float glass, tableware, E-glass fibres and glass wool & fleece. The companies also agreed to head for the 50 % improvement challenge/objective in the whole chain set by the Government.

The Dutch Glass industry faced two major dilemmas at the start

- The investment issue and pay-back time issue: even if opportunities are identified they nevertheless are dependent on the international headquarters regarding investment decisions.
- The competitor issue: As some of the companies are competitors in the market sharing ideas, especially in product development could be limited.

At first a feasibility study was carried out to identify

- the focus areas of improvements for the sector/subsectors;
- the different activities expected to achieve the 50 % target and budget required;
- the commitment of the companies to work on a constructive, focused and realistic but challenging Roadmap.

To carry out the feasibility study as well as the Roadmap the Government provided support to the sectors and companies through Agency NL. Professional consultants were hired and guidance documents were provided to support the companies In identifying the main areas of future innovation.

The feasibility study was carried out with great enthusiasm. As one of the participants indicated this required a lot of "Out of the Bottle" thinking. This resulted in many opportunities for improvement. However, as resources are scarce, priorities had to be set. Selection has not only been based on energy efficiency improvement but also on the expected contribution to sustainability, impact on emissions and economic impact for the companies.

This resulted in 5 focus areas, which are all interrelated to each other.

- Optimisation of production processes, with focus on glass melting;
- Improving product performance and applications;
- Strengthening sustainability of the product chain, with emphasis on recycling and use of energy-extensive raw materials;
- Strengthening the position of glass (products) in society and improving education in glass technology
- Improving innovation power

Two of the focus areas contribute only indirectly to energy efficiency improvement

Improving innovation power is mainly focussed on improving the educational opportunities at various levels in the Netherlands as well as having opportunities for demonstrating new technologies. The basis of innovation is creative, innovative and capable people to be educated at highest possible level .
As an example the role of TNO Glass Group is very important for the Dutch Glass industry. TNO is active in the field of glass melting technology research & education, nationally and internationally,

e.g. with the updated (2011) Glass Technology Training course developed in close cooperation with Dutch glass industry and Agency NL..

Improving the image of glass products and glass applications is mainly aimed at creating a more sustainable image for the Dutch glass industry and its products at consumers, government, students etc. Glass as product should be more visible as modern versatile material in our society.

The other three focus areas are more technically oriented.

Optimisation of processes includes optimisation of the glass composition e.g. new raw materials, batch preparation e.g. pelletizing, improving control of melting process by means of new sensors, new glass furnace designs, design optimization, and optimal use of waste heat from flue gases and other waste streams

Strengthening sustainability of the product chain, with a focus on raising the level of recycling as well as upgrading the quality of recycling material to improve the use of secondary glass or production waste (e.g. filter dusts) and reduce emissions and waste streams.

Improving product performance and applications, by focussing on society trends and future needs given the specific opportunities of glass can identify new applications for glass or glass products. Glass products can contribute strongly to insulation of buildings, energy efficient lighting, light-weight strong construction materials, to solar energy conversion and wind energy turbine efficiencies.

The dilemma's the Dutch Glass industry (investment decisions and competition issues) faced at the beginning of the feasibility study has been reason not to focus on all 5 areas in the process towards a Roadmap. Because of competitive issues Product performance improvement will not be the focus for joint approaches. Strengthening of the social position is not a primary joint activity, since the expected impact in terms of energy efficiency is low.

For the remaining issues it is expected that a 20-40% energy efficiency improvement is still possible for the Dutch Glass industry. The majority of these expected improvements will be in the production process. As the Industry has not yet much experience with quantifying impacts of their efforts in the product chain the expected contribution is rather conservative.

The challenging question posed at the beginning of my presentation will not be answered with a simple YES, but it is clear that a NO does not reflect the view and spirit of the Dutch Glass industry and its innovation possibilities.

The Dutch Glass industry still expects to be able to make major improvements in the near future. By evolutionary and on the longer term revolutionary steps in innovation of production processes and product development. Now, they are investigating which options to focus upon. It is obvious that next steps cannot be made alone. Co-operation with suppliers of raw materials, research institutions, technology suppliers, and government is necessary to achieve the objectives.

As part of the LTA agreement, the Dutch Government will support as much as possible given the actual economic situation. Also a strong international collaboration e.g. through conferences, GlassTrend, joint workshops etcetera, can contribute to achieve the results.

More than 20 years with LTA's and benchmarking showed that cooperation, even among competitors is possible and fruitful. The experiences gained need to be shared, especially internationally. The Dutch glass industry is now identifying new steps for innovation in the future. If you have ideas, please share them with us so we can built upon each other's experience and ideas.

Together one can work on building a more sustainable future. A future using all the benefits the material glass has to offer and that will be different from now.

Experiences gained need to be shared. The Dutch Glass industry and Government is willing to share their experiences in ongoing and new projects and invites others to do the same.

REFERENCES

[1] Novem: Long-Term Agreements on Energy Efficiency, results of LTA1 to year-end 2000 (2001)

[2] Beerkens, R.G.C.; Limpt, van H.A.C.; Jacobs, G.: Energy efficiency benchmarking of glass furnaces. Glass Sci. Technol. **77** (2004) no. 2, pp. 47-57

[3] Beerkens, R.G.C.; Inventory of primary energy consumption and emissions Netherlands glass industry in 2009 (2011), NL Agency project P015610161

Refractories

CONCEPTION OF MODERN GLASS FURNACE REGENERATORS

Stefan Postrach[1], Elias Carillo[1], Mathew Wheeler[2], and Götz Heilemann[3]

[1]RHI Glas GmbH, Germany
[3]RHI Monofrax, USA
[2]Wiesbaden, Germany

ABSTRACT
On one hand, the glass furnace regenerator contributes significantly to the total investments for glass melting facilities. On the other hand they have a major influence on the operating costs of glass furnaces, as only the heat recovery from the exhaust gas enables an energy efficient economic glass production and in consequence competitive product prices. The presentation describes the major factors for the design and operation of modern glass furnace regenerators and discusses their relevance for the refractory material selection.

INTRODUCTION
In the past, glass companies had limited resources and refractory materials that could be used for the design of glass furnace regenerators. Today many glass producers continue to rely on technology that has served them well for many years, but now as glass furnaces become larger, operating conditions more difficult, and capital restrictions more stringent; glass furnaces regenerators are subjected to harsher environments. Companies such as RHI, with a complete line of both refractory materials and expertise in regenerator design can provide complete solutions tailored to every furnace environment. Unlike traditional glass manufacturers, these companies have the experience of seeing many furnaces and many applications as well as access to laboratory testing and furnace modeling that enables them to provide focus and recommendations for each furnace application and situation. The development of new refractory or repurposing of existing materials provides for flexibility in the material selection allowing the refractory manufacturer to zone each area of the regenerator from the crowns and casing to the checker pack for optimum performance. To accomplish this, a partnership is required with the glass producer as no longer can these materials be looked upon as commodities to throw into the regenerator, but instead as high quality products and solutions to provide the best efficiency while maximizing regenerator longevity. With glass furnace regenerators comprising 30-40% of the refractory cost of a new furnace why not take advantage of the in-house expertise of your suppliers when the end result can be a significant reduction in fuel consumption?
At the 2010 Conference on Glass Problems RHI presented our regenerator modeling capabilities, but this paper did not consider potential problems that can occur in the regenerator which can influence both production and efficiency. By selecting an adequate refractory lining, these problems can be minimized. This paper will take a more simplistic, but not less technical look at specific applications within the regenerator and the material solutions that can be supplied for each of the furnace conditions that may be present.

OPERATING CONDITIONS
The focus of this paper and thus the operating conditions discussed will be related to end-fired furnaces. End-fired furnaces tend to exhibit the higher probability to have problems primarily due to their higher specific load/m[2] and the strong carryover effects inherent in their design. In contrast, side-fired furnaces have more limited issues and many of the design practices discussed in end-fired applications can also be translated to side-fired designs.

Upper Chamber Area

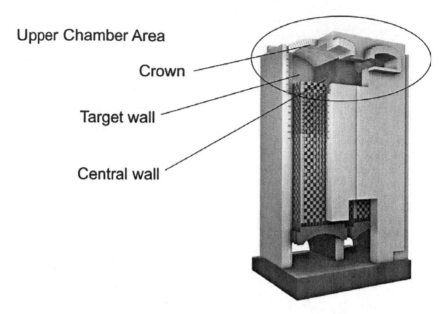

Crown

Target wall

Central wall

Figure 1. Section view of a typical end-fired furnace design.

The operating conditions of an end-fired furnace can be broken down into three basic considerations: temperature, mechanical load, and chemical attack. These areas must be taken into account together and separately in the design of the regenerator casing and also be kept in mind in for the design of the checker pack. Unfortunately this information is often not supplied or may not be available from the glass producer during the material specification process.

Temperature:
Temperature is a critical factor in the selection of refractory materials for the regenerator, especially for the crown and upper walls with special emphasis on the center or division wall. With the development of improved burner systems waste gas inlet temperatures are lower than they were in the past; often reported as approximately 1400°C, with areas at the top of the regenerator as high a 1550°C. This is a stark contrast from temperatures reported in 1984 when waste gas inlet temperatures greater than 1500°C were typical and the temperature at the top of the regenerator was often up to 1600°C [1]. Unfortunately, the distribution of the temperature in the regenerator is mostly unknown as only recently has temperature modeling become available and even with this availability the regenerator has been largely neglected. Possibly this neglect is due to the difficulty inherent with developing a model that can accurately account for all variables occurring within the regenerator and checkers.

We often focus on temperature and its distribution within the regenerator crown as being important, but this can be controlled with crown design and insulating practices. In fact most regenerator crowns could be built with slightly lower grade refractories if not for the concern for temperature excursions and the desire to insulate heavily for heat saving purposes. In reality the most crucial application of refractory when considering temperature is the central wall due to the high mechanical load in traditional construction and nearly no temperature gradient as it does not have a cold side. According to RHI calculations the temperature difference in the central wall during the firing and exhaust cycles is less than 150°C. Improper selection of center wall material can lead to serious complications.

Mechanical Load:

The modern glass furnace trend is to go to larger furnaces where today 450 tons/day is no longer a rarity and a furnace with a melter area of 175m^2 (~1883ft^2) has even been announced. In order to compensate for the waste gas volume from furnaces of this size the checker volume needs to become larger, but the checker height is typically limited to 12m (~39 ft) resulting in the need for a larger cross sectional area with greater than 30m^2 (323 ft^2) becoming common. These larger regenerators lead to additional load on the regenerator crowns and central walls with outer walls being less of a concern. The question that remains is whether these larger regenerators are efficient. Is uniform perfusion achieved in the larger regenerators? Do the furnace design companies need to consider new or even possibly old design ideas related to regenerator size shape and exhaust flues?

Chemical Attack:

Chemical attack in the regenerator is normally related to side effects of combustion fuels, batch constituents (carry-over), and waste gas vapors. Often this attack is enhanced by the glass producers' efforts to reduce energy costs. Several mechanisms can be related to these efforts:

- Cheap Oil → Attack from high V_2O_5 and S contents
- Batch pre-heater → dry batch, increased carryover
- Fine sand → increased carryover
- Cheap dolomite → blow up tendency
- High cullet ratio → impure cullet quality and high amount of fines
- Initial attack by blowing cullet during heat up process → crown, walls are glazing

The primary effects of the above attack mechanisms are seen in the target walls, crowns, and top checker layers.

To summarize the operating conditions, in today's era of modern glass furnace design many of the operating conditions found in the upper regenerator area are not available or are unknown. Mechanical load at high temperature must be considered, but exactly what is the definition of mechanical load and high temperature? Many methods of decreasing fuel costs can lead to different attack mechanisms in the regenerator. These unanswered operation questions make adequate refractory selection in the regenerator difficult, especially under today's economic constraints where the highest quality and thus highest cost materials are not always an option. Open communication and partnership between the glass producer, furnace designers, and refractory suppliers are essential to ensure long term success.

MATERIAL SOLUTIONS

During the 2006, 67th Conference on Glass Problems, RHI presented a paper that discussed in detail the material selection for checker packs, therefore checker material and selection parameters will not be presented in detail in this paper. Instead, this paper will focus on the material requirements and solutions for the regenerator casing and later discuss checker design and efficiency. It is worth noting that the new spinel bonded magnesia product introduced in the 2006 paper has completed initial trials and testing; proving to be a viable alternative to the traditional magnesia-zircon used in the condensate zone of checker packs operating under oxidizing conditions. With the current rapid inflation of zircon raw material prices, the spinel bonded magnesia brick provides glass producers an option to keep their refractory costs under control in this area of the furnace.

Table I. Characteristics of the spinel bonded magnesia brick.

Chemical composition [wt. %]				
SiO₂	Al₂O₃	MgO	CaO	Fe₂O₃
1.4	23.6	73.6	0.9	0.5

Physical characteristics				
Bulk Density [g/cm³]	Open porosity [vol.%]	Cold crushing strength [MPa]	Thermal expansion [%]	Refractoriness under load [T₀.₅]
2.95	15.0	60	1.4 (1400 °C)	1590 °C

Figure 2. Spinel bonded MgO after 3 years service in the condensation zone of a soda lime container glass furnace.

The proper material selection for the regenerator casing (walls and crowns) is vital in achieving the long term performance desired by most glass producers. Before the application of the potential material solutions is explored, as with understanding operating conditions, the available materials themselves should be presented. As discussed with the operating conditions, all aspects of the regenerator operations must be understood and communicated between companies to ensure that the proper material selection is made.

Silica Products:
Prior to the use of magnesia and later higher alumina materials in the regenerator, silica products were the workhorse of the glass industry. Later the introduction of 'no lime silica', originally designed for oxy-fuel crown applications, has provided new options for the regenerator casing

application. These products bring many excellent properties to the table, but also have their limitations, which have lead many away from using silica.

Major characteristics of Standard Silica for regenerator applications:
- SiO_2 content approximately 96%; bonding is $CaOxSiO_2$ (wollastonite)
- Excellent hot properties to approximately 1600°C (2912°F)
- Good thermal expansion and low thermal conductivity
- High resistance against SiO_2 attack even against fine silica sand
- V_2O_5 resistance is limited due to CaO content
- Low resistance to alkali and cullet
- Low resistance at T < 1450°C due to tridymite

Major characteristics of 'no lime silica' for regenerator applications:
- SiO_2 content is approximately 98%; bonding is w/o $CaOxSiO_2$ (wollastonite)
- Excellent hot properties up to 1650°C
- Excellent thermal expansion and low thermal conductivity
- High resistance against SiO_2 attack even against fine silica sand
- Good V_2O_5 resistance due CaO-free bonding
- Improved resistance against alkali
- Good corrosion resistance at T < 1450°C due to limited tridymite content

High Alumina Products:
Andalusite and mullite high alumina based materials have always been options for regenerator construction, but only recently have they gained notoriety as the newest solution for regenerator problems. Perhaps this can be attributed to changing operating conditions or maybe it is due negative press related to problems with magnesia in isolated cases. Nevertheless, like any refractory material, care must be taken when selecting andalusite and mullite materials as there are a wide range of andalusite/mullite grades available with each exhibiting different properties.

Andalusite reaction:
- Andalusite is a natural raw material Al_2SiO_5
- During firing, starting at approximately 1200°C, conversion to mullite and SiO_2 rich glassy phase takes place
  - $3 Al_2SiO_5 = 3 Al_2O_3 \cdot 2 SiO_2$ (mullite) $+ SiO_2$
- To achieve 100% mullite additional Al_2O_3 is necessary:
  - $3 Al_2SiO_5 + Al_2O_3 = 3 Al_2O_3 \cdot 2 SiO_2$ (mullite)
- Volume expansion of the reaction is 3-5%
  - Attention: If the conversion to mullite is not complete during firing, the expansion may overlap with creep distorting the test results.

Major characteristics of andalusite for regenerator applications:
- Al_2O_3 content between 57% (pure andalusite) and 72% (100% mullite 3:2)
- 77% Al_2O_3 grades are available
- Refractoriness under load: drastic drop at T>1500 for 64% Al_2O_3
  - Critical for upper wall, central wall, and crown applications

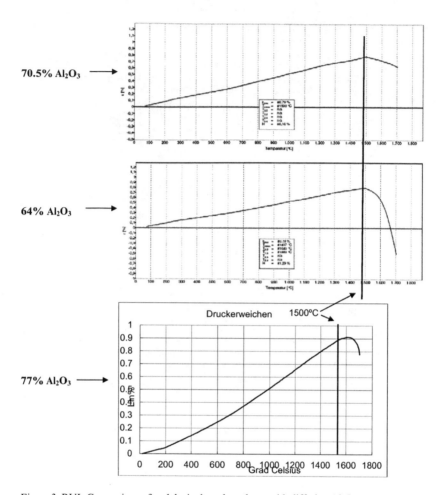

Figure 3. RUL Comparison of andalusite based products with differing Al₂O₃ contents.

- Improved creep resistance due to adapted production parameters and raw materials (e.g. fused corundum addition)
  - T_0 temperature for 70.5% Al_2O_3: 1500°C
  - T_0 temperature for 77.0% Al_2O_3 grade: 1610°C
- High resistance to SiO_2 attack (even fine sand) and V_2O_5 attack
- Alkali resistance attributed to glassy phase reaction layer (prevents alkali infiltration
  - Glassy phase also buffers nephaline formation (limited spalling)

    o Long term stability of the protection layer, especially under changing operating conditions remains a question mark
- Resistance against cullet is limited
- Low thermal conductivity and thermal expansion

Major characteristics of fused mullite for regenerator applications:
- Raw material is fused mullite (higher Al_2O_3 content due to 2:1 mullite); Al_2O_3 content is approximately 76%
- Firing at $T > 1700°C$
- Excellent creep resistance even at 1650°C

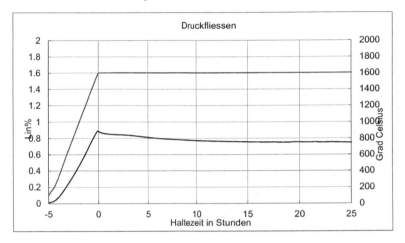

Figure 4. Creep resistance of fused mullite at 1650°C

- High resistance to SiO_2 attack (even fine sand) and V_2O_5 attack
- Higher Al_2O_3 critical in case of alkali attack (fused mullite is less affected than sintered
- Nepheline formation at lower temperatures and spalling possible (due to absence of glassy phase that buffers resulting stresses!)
- Resistance against cullet is limited
- Low thermal conductivity and thermal expansion

       Alkali vapor testing of various andalusite/mullite grades was conducted to determine the depth of penetration of alkali into the test sample. As discussed in the product characteristics, the depth of penetration can vary depending on the material composition of this family of refractories. The results shown in Figure 5 confirm that the glassy phase present in the andalusite grades can limit alkali penetration where as the fused mullite materials tested both show greater amounts of alkali penetration. However, the question remains; do the changing operating conditions of the regenerator affect the long term stability of the glassy phase protection layer?

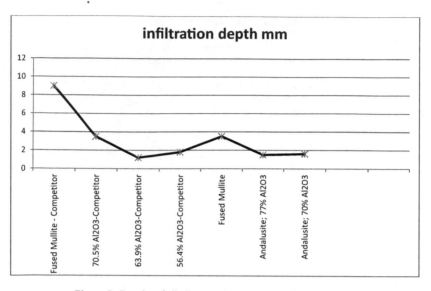

Figure 5. Results of alkali vapor test showing infiltration depth.

Magnesia Based Products:

Like the high alumina refractory grades, there are many different grades available for the magnesia family. Important variables in the manufacture and selection of these materials are raw material quality, final chemistry of the refractory, and firing temperature. Together these factors determine the final properties of the material and thus allow them to be zoned into the proper areas of the regenerator. For the purposes of this paper, the characteristics discussed below apply primarily to materials for the upper structure of the regenerator as this is where the majority of the problems in the casing can occur.

Major characteristics of magnesia for regenerator applications:
- ☒ Adequate raw material consists of coarse crystal grains (<150μm crystal size; fused or sintered
- ☒ MgO content is 97% / CaO:SiO$_2$ ratio > 2.2
- ☒ High firing temperature → high degree of direct bonding
- ☒ Excellent creep behavior at 1600°C
- ☒ Excellent behavior against alkalis → high stability under reducing conditions
- ☒ Reaction with SiO$_2$ to forsterite → spalling especially at the beginning of a furnace campaign
- ☒ Higher thermal expansion → has to be controlled more carefully
- ☒ Higher thermal conductivity → requires adapted insulation

SUMMARY: MATERIAL SOLUTIONS

In summary, operating conditions will often dictate the optimum refractory material for each application. For example high alumina products may be an excellent choice for furnaces that utilize fine sand as their source for silica. Conversely if there is high alkali vapor or carryover content or fine cullet being used, high magnesia materials are more likely the better choice. Zoning a regenerator to take advantage of lower cost material options is a valid design solution, but understanding both the materials and operating conditions such as temperature profiles within the regenerator, mechanical load and chemical attack are critical to avoid using materials that may not withstand present conditions or potential excursions outside of normal operating conditions. Table provides a brief overview of the advantages and limitations of the described materials.

Material	Advantages and limitations
Silica	• Silica in the regenerator crown is cost efficient and suitable for situation where there is significant sand carry-over and high flue gas temperatures (>1470°C or 2678°F). The formation of liquid glassy phase due to the attack or alkalis is critical in the case of lower flue gas temperatures
	• Experience has show that a silica crown combined with magnesia walls should be avoided because silica melt run-down form the crown corrodes the magnesia bricks
	• An alternative to standard silica is silica bricks without lime (e.g. STELLA GNL)
High Alumina	• In the case of very strong sand carry-over (mainly SiO_2) and a lower alkali content in the flue gas, mullite bricks (e.g. DURITAL E75EXTRA and DURITAL S70) can be used
Magnesia	• Magnesia bricks have a high resistance against alkali attack. For high temperature and crown applications a magnesia grade with low creep value (e.g. ANKER DG10 and RADEX SG-CN) is recommended
	• A regenerator casing comprising a magnesia crown and walls has the best cost-performance ratio of all the different material concepts

Table II. Overview of the refractory material advantages and limitations for regenerator casing applications.

DISCUSSION MAGNESIA OR HIGH ALUMINA PRODUCTS:

In recent years the comparison of magnesia to high alumina or mullite materials for regenerator casings has expanded the use of mullite materials for regenerator casings. As a manufacturer of the full range of both magnesia and high alumina materials RHI maintains an open mind when considering all material solutions. Each category does have their unique advantages that can be utilized in different operational situations. The fact remains that all aspects of the operation must be understood and all material options considered carefully when selecting the proper refractory grades. As previous publications have covered the properties inherent in mullite the following information will attempt to answer some questions that have been raised regarding the use of magnesia.

History:

Magnesia materials have enjoyed an excellent long term history in serving the glass industry. In fact RHI alone has installed more than 130,000 metric tons in over 600 furnaces worldwide. Campaigns of 20 years and more are not unheard of when utilizing magnesia materials. As with high alumina materials the selection of the proper magnesia based refractory is critical to the safe and long term operation of the furnace. While there have been failures of regenerators utilizing magnesia how many of these failures can really be attributed to cases when the correct as well as the highest quality material was selected for the application?

Price:

High quality, high magnesia materials are available at prices that are quite competitive when compared to high alumina based materials. In fact fused mullite grades are not inexpensive. Like mullite regenerators, magnesia regenerators can be zoned to utilize the highest grades where needed in critical applications and then change to more cost effective, lower grade materials in middle walls and below to take advantage of cost savings available.

Thermal Expansion:

Magnesia materials do have significantly higher expansion rates than high alumina materials such as andalusite and mullite. This can be a problem if not considered and dealt with using proper furnace design and consistent operation practices during initial startup. Due to the size of the regenerators involved, expansion allowance and control is certainly more of an issue in a side-fired float furnace than the traditional end-fired container furnace and may really be the reason float glass producers have been looking to alternative materials. Mullite materials are more forgiving in this area, not requiring the special attention that magnesia materials do, but traditional construction techniques often used in other industries such as cardboard spacers, key bricks, and others can help minimize potential problems with expansion in magnesia.

Thermal Conductivity:

Potential energy savings of millions of dollars over the life of the furnace has been theorized in recent publications [2]. While it is true that magnesia materials have significantly higher thermal conductivities than mullite materials, several recent publications have indicated that the actual total heat losses from modern, highly efficient regenerative furnace are quite low. Figure 6 below represents a typical Sankey diagram for a modern regenerative container glass furnace. From this diagram it can be interpreted that the share of regenerator heat losses is below 3% of the total energy input to a fuel-fired glass furnace. Therefore, any change in the lining layout in the outer upper structure would hardly achieve substantial energy savings beyond 3 % in the long term despite the high temperatures in that region.

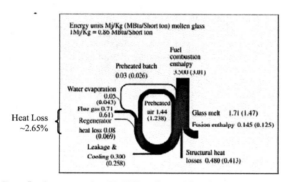

Crossfired regenerative 70-75% cullet and batch preheat Sankey diagram from Ruud Beerkens, TNO; represents one of the 10% most energy efficient furnaces (container glass) in Europe

Figure 6. Typical Sankey Diagram for a modern regenerative container glass furnace [4].

Furthermore, the use of a proper insulation package can significantly minimize these differences. In fact, the insulation package is the key to saving energy and thus the thermal conductivity of magnesia products has not been considered a significant issue at RHI. Proper design and insulation practice can bring nearly as much energy savings to the regenerator as changing from magnesia to mullite in the walls and crowns. The idea of thinning the wall using mullite has also been presented [2]. This involved changing the hot face wall thickness from 375mm to 250mm and increasing the insulation package. While this indeed is a good proposal, it is not a new idea to magnesia users. The same practice is already used in the construction of magnesia regenerator walls. Only in the target wall is the recommended wall thickness 375mm. Most glass producers can and do use 250mm outer walls in their regenerator design with 375mm tie-back courses. Often behind the hot face is a 375 mm back up layer of IFB grades ending up with a 25 mm gunned insulation grit mix. The list below is an example of a typical lining from hot to cold face.

Typical Magnesia Wall Lining Example:
250 mm Inner lining: 97% MgO (with 375mm tie-back)
125mm Insulating MgO between tie brick layers
125 mm IFB, type: 140-0,80-L (2600ºF)
125 mm IFB, type: 1250-0,50-L (2300ºF)
125 mm IFB: type: 125-0,50-L (2300ºF)
25 mm Vermiculite insulation mix.

As presented during the 2010 Conference on Glass Problems [3] using a dynamic approach to the heat transfer model rather than a steady state condition provides a more complete picture of heat loss from the regenerator. An example of the dynamic heat transfer model is shown in Figure 7 which presents a view of the magnesia and mullite concepts taken as a snapshot 10 minutes into the reversal cycle. Using this type of model, several heat transfer parameters such as thermal conductivity and heat loss can be calculated during a unit of time in both the hot and cold reversal cycles. Viewing in this manner, one can see in Figure 7 a side by side comparison of the magnesia and mullite regenerator wall concept of both the hot and cold cycles. As discussed above the key to managing the heat loss is insulation. Adding as little as 85mm of additional insulation to the magnesia design equalizes the difference between thermal conductivities of the magnesia and mullite materials. The net result, seen in Figure 8, is nearly a negligible difference in heat lost between a magnesia and mullite wall of similar construction.

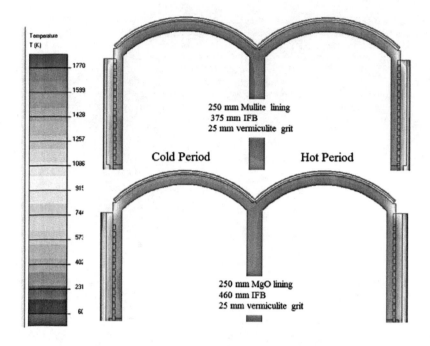

Figure 7. Dynamic view of heat transfer in an MgO and Mullite walls

MgO (460 mm IFB)				Mullite (375 mm IFB)		
Cold period		Hot Period		Cold period		Hot Period
W/m2	minute	W/m2		W/m2	minute	W/m2
831.5	0	849.37		812.6	0	843.92
820.29	1	836.56		805.89	1	845.27
812.63	2	827.87		800.83	2	846.72
807.19	3	821.82		796.75	3	848.2
803.23	4	817.48		793.29	4	851.44
800.16	5	814.32		790.21	5	853.19
797.69	6	811.84		787.4	6	855.02
795.59	7	809.82		784.76	7	856.94
793.74	8	808.1		782.26	8	858.95
792.03	9	806.57		779.86	9	861.01
790.46	10	805.2		777.57	10	863.22
789	11	803.87		775.36	11	865.53
787.54	12	802.62		773.25	12	867.97
786.15	13	801.47		771.2	13	870.56
784.79	14	800.36		769.22	14	873.36
783.48	15	799.29		765.49	15	876.43
782.17	16	798.25		767.32	16	879.91
780.92	17	797.24		761.05	17	882.97
779.69	18	796.26		762.03	18	884.04
778.47	19	795.3		760.44	19	889.22
777.28	20	794.36		758.89	20	896.12
15869.61	mean	16176.105		15589.93	mean	17299.97
793.48	value	808.81		779.49625	value	865.00
801.14			-3%	822.25		

Net Heat Loss

Figure 8. Comparison of net heat losses of MgO/Mullite concepts.

Finally, to answer the remaining question regarding utilizing the same insulation package for the magnesia and mullite concepts, Figure 9 illustrates that when using 375mm of insulation for both models, there is indeed a difference in the net heat loss. This difference, approximately 18%, may appear to be significant, but when referring back to the previously discussed Sankey diagram (Figure 6) one will note that the maximum share of heat loss for the glass melting operation from the regenerator walls was less than 3%. Therefore the maximum savings that may be achieved by switching from magnesia to mullite materials is only approximately 0.5%. Considering that the discussion so far has only been related to the high temperature upper layers of the regenerator, the contribution of this area actually to the total heat loss is not the entire 3% making the savings from changing materials even lower. Consequently a furnace designer/operator working to improve the efficiency of their regenerator may be better off spending their time evaluating checker concepts and operating parameters than making changes to existing materials.

Mullite (375 mm IFB)				MgO (375 mm IFB)		
Cold period		Hot Period		Cold period		Hot Period
W/m2	minute°	W/m2		W/m2	minute	W/m2
812.6	0	843.92		984.76	0	967.09
805.89	1	845.27		975.5	1	968.52
800.83	2	846.72		969.15	2	969.99
796.75	3	848.2		964.49	3	971.49
793.29	4	851.44		960.83	4	973.04
790.21	5	853.19		957.79	5	974.62
787.4	6	855.02		955.11	6	976.24
784.76	7	856.94		952.67	7	977.91
782.26	8	858.95		950.38	8	979.62
779.86	9	861.01		946.08	9	981.38
777.57	10	863.22		944.03	10	983.29
775.36	11	865.53		942.02	11	989.48
773.25	12	867.97		940.06	12	987.17
771.2	13	870.56		938.14	13	989.29
769.22	14	873.36		937.39	14	991.61
765.49	15	876.43		936.25	15	994.24
767.32	16	879.91		934.4	16	997.34
761.05	17	882.97		932.59	17	1001.2
762.03	18	884.04		930.82	18	1006.3
760.44	19	889.22		929.07	19	1013.5
758.89	20	896.12		927.37	20	1024.1
15589.93	mean	17299.97		18952.835	mean	19721.825
779.49625	value	865.00		947.64	value	986.09125
	822.25		18%		966.87	

Figure 9. Comparison of net heat losses of MgO/Mullite concepts utilizing the same insulating design.

CHECKER PACK:

No modern regenerator paper would be complete without a short statement regarding the checker pack design. As mentioned, previous publications from RHI discussed checker material selection. As an addition to this, some points should be considered in the selection of checker block design.

Checker block selection is always a compromise between efficiency and clogging. Higher efficiency is achieved using checker brick that have smaller flues and add turbulence. Lower clogging tendency is realized with larger flues and no turbulence. An optimum pack design must be identified. The modern glass furnace regenerator uses either thin walled chimney blocks or cruciform checkers. Both designs offer shapes with different flue sizes and that introduce turbulence into the regenerator. Where these shapes are used may affect the life and long term performance of the regenerator.

- Top Layers: Turbulence accelerates corrosion by batch carryover
  → Top checker should not create turbulence
- Hot Zone: No significant corrosion → Main area of heat exchange
  → Checker should create turbulences
- Condensation Zone: Turbulence accelerates corrosion by sulphates
  → Checker should not create turbulences

In addition to the checker geometry, material characteristics such as corrosion resistance and glassy phases can influence the clogging behavior and therefore the efficiency. Proper material selection must be considered together with the optimum checker selection.

CONCLUSIONS:

In conclusion the modern glass furnace regenerator remains an increasingly important subject for the glass manufacturer. With the many materials and design options available to the furnace designers and operators, a good understanding of all operation aspects is critical to the refractory supplier to make proper recommendations. While changed operating conditions have made andalusite and mullite materials more popular for construction of regenerators, high quality magnesia continues to have excellent performance history and references especially in U-Flame end-fired furnaces, even operating under severe conditions. With the wide range of characteristics inherent in the different families of refractories available for regenerator construction, careful selection of refractory as well as a high quality, knowledgeable supplier is necessary.

REFERENCES
1. Glass Furnaces: Design, Construction and Operation, W. Trier., The Society of Glass Technology, 1983.
2. "Mullite Regenerators – An Optimum Solution", C. Windle and T. Wilson, DSF Refractories and Minerals Limited, 70th Conference on Glass Problems, Charles H. Drummond, III, Editor, 2009.
3. "Regenerator Temperature Modeling for Proper refractory Selection", E. Carillo and M. Wheeler, RHI Refractories, 71 Conference on Glass Problems, Charles H. Drummond, III, Editor, 2010.
4.Glass Making Technology: a Technology and economic assessment: C.P. Ross & G.L. Tincher, 2004.
http://www.osti.gov/glass/Special%20Reports/Glass%20melting%20tech%20assessment.pdf

NEW CRUCIFORM SOLUTIONS TO UPGRADE YOUR REGENERATOR

D Lechevalier[2], I Cabodi[1], O Citti[2], M Gaubil[1], J Poiret[3]

[1] Saint-Gobain CREE, Cavaillon, France
[2] Saint-Gobain NRDC, Northborough MA, USA
[3] Saint-Gobain SEFPro, Le Pontet, France

ABSTRACT

SEFPRO will present in this paper new solutions for glass furnace regenerator adapted to the trends of glass industry towards friendly environmental running conditions. To face energy cost increase and CO_2 emission, SEFPRO developed new cruciform design to improve significantly thermal efficiency of packing with higher heat exchange rate. New refractory composition for regenerator has also been developed to follow glass furnace running conditions toward low NOx emission. These new solutions will be discussed in terms of technical performance and also of economical and lasting life benefit for glass industry.

CONTEXT

Energy consumption reduction is more and more concerning the glass industry. Representing up to 20% of the glass production cost, the share of energy is constantly increasing despite numerous glass melting process optimizations. In addition to the production cost, energy consumption drives the emission of heavily taxed NOx and SOx emissions.

As a requirement for the glass industry competitiveness and its environmental friendly orientation, this overall trend induces new solicitations and expectations for the refractory used in the checker pack.

Energy consumption reduction requires an optimization of the thermal exchange in the checker pack to go a step forward checker design optimization. Therefore shapes of the refractory, as a key driver to increase checker pack efficiency, as to be adapted to the various type of heat transfer occurring in a regenerator.

Reduction of NOx and SOx emissions may be achieved by the use of techniques such as primary settings that leads to reducing atmosphere (meaning CO in excess) in glass furnaces. This modification of the furnace atmosphere equilibrium changes the composition of the condensates that appear in the checker pack and increase the solicitation on the refractory materials that are used.

NEW SHAPE TO INCREASE THERMAL EFFICIENCY

Regenerative furnaces are a very competitive solution to produce glass economically. However the regenerator chamber performance is not yet maximized despite the numerous improvements in checker designs over the years. SEPR has studied the specific phenomena encountered in glass furnace regenerator in a dual approach: experimental tests in a high temperature large scale experimental setup to verify the concepts, in parallel with specific numerical models to study a wide range of configuration and reliably predict the performance of industrial regenerators. As a result, a complete range of checkers with different materials and shapes is proposed, enabling the glass maker to adapt the regenerator packing to the constraints of the glass furnaces: corrugated checker[i] (Type 4) for enhanced heat transfer, large flue size[ii] (Type 6) to have modular solution in the lower courses, higher

91

corrosion resistant material (ER5312RX) for top course checkers and more recently ER55XX material that is adapted to reducing conditions. Our experimental and numerical tools are also able to demonstrate the superiority of such cruciform designs over other available designs. The latest improvement that is introduced in this paper is the checker packing with a new Type 8 cruciform.

Figure 1 : Type 3 (smooth, *left*) and Type 4 (corrugated, *center*) and new Type 8 (*right*)

Although the theoretical energy efficiency of a regenerator lies around 75 to 80%, depending on parameters such as the type of fuel or the combustion settings, the actual performances cannot in practice reach the maximal efficiency. Type 8 allows to go beyond the usual limits and getting closer to the theoretical value while keeping the regenerator chamber dimensions as compact as possible. This paper presents the technical benefits of this new solution and the results of numerical models and experiments, as well as its industrial validation.

Numerical Modeling

The heat regeneration theory has been described extensively by different authors[iii][iv]. The thermal regenerator is described by the coupled energy balance equations of the gases (alternating between air and waste gases) and the energy balance of the checkerworks. The convective heat transfer coefficient is determined by the local flow regime, which, depending of the design of the regenerator and the local conditions within the packing, could be a mix of forced and natural convective flow regimes. These different regimes have been reproduced by a specific regenerator CFD model with ANSYS-Fluent® to take into account the physics of the problem for a single channel: gas radiation, effect of transition between different checker designs, turbulent convective heat transfer, and with a specific development to account for natural convection contribution in vertical channels that has been described in a previous publication[v]. The Fluent® based regenerator model solves a transient calculation that alternates between a period during which air flows upwards, and a period in which flue gases flow downwards. The convergence is reached when at each checker level the energy released to the air and the energy gained from the flue gases are equal. The model reproduces well the different flow regimes encountered in a glass furnace regenerator as illustrated in figure 2 in the case of a large regenerator (0.20Nm/s in air, 13m high).

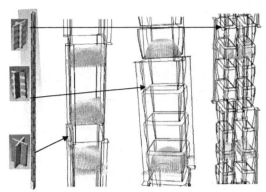

Figure 2 : Predominant flow regimes in a regenerator channel during air period: Natural convection (Type 3, *left*); Turbulent forced convection (Type 4, *center*; Type 8, *right*)

A second model based on the detailed observation of CFD results as well as experimental data has been developed. In this simplified formulation, the description of the physics is retained in the energy balance of the 3 zones of the problem that are air, waste gases and checkerworks. Heat exchange between these 3 zones is handled through wall to wall radiation, gas – solid radiation, and convection. For a single channel of specific surface S_p (expressed in m²/m³), the local equations for air, waste gases (*wg*) and checker matrix can be written at a given altitude *z* as follow:

$$\rho_{air}Cp_{air}dV_{gas}(z)\frac{dT_{air}}{dt} + \dot{m}_{air}Cp_{air}\frac{dT_{air}}{dz} = -h_{air}.Sp.(T_{air} - T_{checker}) - \phi^{rad}_{checker} - \phi^{rad}_{air-checker}$$

$$\rho_{wg}Cp_{wg}dV_{gas}(z)\frac{dT_{wg}}{dt} + \dot{m}_{wg}Cp_{wg}\frac{dT_{wg}}{dz} = -h_{wg}.Sp.(T_{wg} - T_{checker}) - \phi^{rad}_{checker} - \phi^{rad}_{wg-checker}$$

$$\rho_{checker}Cp_{checker}dV_{checker}(z)\frac{dT_{checker}}{dt} = h_{gas}.Sp.(T_{gas} - T_{checker}) + \phi^{rad}_{checker} + \phi^{rad}_{gas-checker}$$

The contribution for the wall to wall radiation is calculated by view factors between each altitude *z* in the packing. The gas radiation contribution uses the gas emissivity calculated from Hottel tables with the Leckner correction[vi]. As a major improvement to previous work, the convective contribution is described by a local heat transfer coefficient in air period and in waste gas period as $h_{gas}(z) = h_{air}(z)$ of $h_{wg}(z)$. An empirical diagram for the regime determination in the simplified case of a smooth channel has been proposed by Métais and Eckert[vii] and has been adapted for different checker designs for glass furnace regenerator.

This model can predict accurately the heat transfer phenomena in the specific conditions of a glass furnace regenerator at an affordable computation cost. This model has been validated against experimental data and is now in use for customer support at SEFPRO.

The following sections present the experimental and the numerical validation of the concept of regenerator with the new checker described above. In this paper the energy efficiency η_{regen} is the ratio of the energy recovered by air E_{air} during a period by the available energy in the flue gases E^{max}_{flue}. The definition of these energy quantities is given by the following relations:

$$E_{air} = \int_\tau\int_{Tair_{inlet}}^{Tair_{outlet}} \dot{m}_{air}Cp_{air}dTdt \; ; \; E^{max}_{flue} = \int_\tau\int_{Tair_{inlet}}^{Tflue_{inlet}} \dot{m}_{flue}Cp_{flue}dTdt \; ; \qquad \eta_{regen} = \frac{E_{air}}{E^{max}_{flue}},$$

where $Tflue_{inlet}$ is the inlet temperature of the flue gases in the regenerator, $Tair_{inlet}$ is the inlet air temperature and $Tair_{outlet}$ is the air preheating temperature exiting the regenerator.

Experimental Set up

The experimental set up used for the regenerator study is based in CEA-GRETh laboratory in Grenoble, France. It has been described in previous publications[viii] and its principle is shown on the sketch of figure 3. The set up can test two regenerator packing configurations simultaneously in two chambers and in the same flow rate and temperature conditions, close to an actual glass furnace working point. The test chamber has a usable height of 5m and a 6x6 channels cross section of 150x150mm flues. The gas temperature is measured at the top and the bottom section with aspiration pyrometers. The solid temperature is monitored by thermocouples placed in the checker in the central channel.

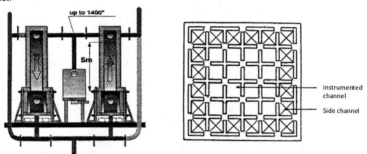

Figure 3 : Principle of the laboratory test loop with 2 regenerator chambers (*left*)
and the details of the chamber cross section (*right*)

For the purpose of the present study we have compared two packing geometries: the reference that consists in the combination of smooth cruciform (Type 3) in the lower section and corrugated cruciform (Type 4) in the upper section, and an example of the innovative packing solution with Type 8 made in this case of the combination, from bottom to top sections, of Type 6, Type 3, Type 4 and Type 8. The comparison between these packing designs is summarized in the table 1 as well as the results of the measurements.

	Reference Types 3-4	New Solution Types 6-3-4-8
Weight (tons)	5.17	5.06
Surface area (m²)	17.2	18.7
Air flow (Nm³/h)	866	845
Fumes flow (Nm³/h)	1205	1205
Air inlet (°C)	153	155
Air outlet (°C)	934	1014 (+80)
Flue gas inlet (°C)	1179	1151
Flue gas outlet (°C)	658	571
Energy Efficiency	44%	50% (+13.6%)

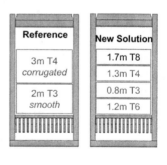

Table 1 : Packing descriptions and laboratory results
based on gas temperature measurements

For very similar working conditions (gas flow rates and inlet temperatures) the innovative packing preheats the air 80°C higher than the reference and therefore the energy recovery is almost 14% higher. It is also important to note that this performance increase is achieved in this particular case at a similar total packing mass, which prefigures the fact that the solution could be very economical.

Furthermore, the laboratory set up gives insight on how the regenerator works. The local heat exchange intensity can indeed be derived from the checker temperature measured at different levels along the height of the central channel of the packing. The local exchanged power, figure 4, shows that the increase of performance is due to a higher heat exchange in the top section where the Type 8 cruciform checkers have been installed.

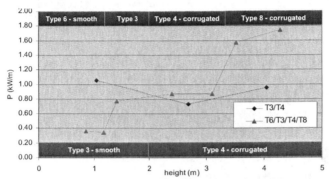

Figure 4 : Exchanged power (in kW/m) vs packing height in the laboratory set up

Industrial Validation

In order to validate the concept, we need to find a furnace on which it would be possible to compare a pair of chambers with Type 8 and a pair of chambers without Type 8. For an end fired furnace, there is only one pair of chamber, hence the comparison can only be done between the situations before and after a cold repair. The furnace would need to be set to equivalent working conditions (pull rate, temperatures), which is difficult to obtain in the industrial reality. Furthermore it is well known that furnace efficiency is higher at the beginning of a campaign so to carry out a fair comparison we would have had to find two similar furnaces that were repaired at the same time and working in the same conditions. Such opportunity is of course very unlikely, not to say impossible. The alternative was to test the solution on a cross fired furnace and to insure that two pairs of chambers would work as close as possible in the same conditions (gas consumption, air excess and inlet temperature levels). Again in the industrial world it is almost never the case. However we found a good candidate and two pairs of chambers were installed for a side by side comparison. These two chambers were engineered to work at the same fuel consumption. Based on the model presented earlier, the comparison between the solution without Type 8 and the solution with Type 8 gives an increase of energy efficiency over 3.5% (table 2). Two cases have been considered: Type 8 is substituted to Type 4 over 23% of the height of the regenerator in the first case, and 27% in the second case. The improvement between these two cases was quite marginal, therefore is has been chosen to conduct the trial with the top 23% of the chamber to be installed with Type 8 for the tested configuration.

Calculated heat balance (MWh/channel)	Standard Type 3 – Type 4	Type 8 – 23% of the height	Type 8 - 27% of the height
Available Energy (nominal assumption)	9.13		
Regenerated Energy	5.90	6.22	6.27
Energy efficiency	64.7%	68.2%	68.8%

Table 2 : Calculation comparison between standard and new configuration cases based on nominal conditions of the furnace (Regeneration Fluent® model)

Similarly to the laboratory set up, we were able to evaluate the performance of the packing in two ways. First, the heat balance of the complete chamber is based on the flow rates and the temperature at the top and bottom. The integrated energy values are summarized in the table 3. It is almost impossible to get the same exact working conditions between two pair of chambers. This being stated, the different measurements done on the furnace gave quite similar flow rates for air and for flue gases in the two test chambers. The inlet temperature of the air and the flue gases was slightly different and that explains the difference in available energy in the two chambers. The resulting energy efficiency is greatly in favor of the solution with Type 8. It confirms the trend of the lab scale results as well as the order of magnitude that is predicted by the numerical model.

Measured Heat Balance (MWh/channel)	Standard Type 3 – Type 4	With Type 8
Available Energy	12.45	11.89
Regenerated Energy	8.58	8.90
Energy efficiency	69.0%	74.8%

Table 3 : Measured heat balance of the two test chambers

Second, the instrumentation of the checkers was carried out for the two test chambers of the furnace. One central channel was equipped during the construction of the chambers with thermocouples at different height levels. The thermocouples are located in the wing of a specially machined cruciform and connected to the outside of the chamber through a dedicated opening. The cables are protected inside a specific groove (in red in figure 5) between two cruciform layers in order to minimize their disturbance to the furnace operation. The thermocouples survived the start up of the furnace and were working well after the furnace conditions and pull rate were stabilized so that relevant temperature profiles can be derived from the measurements.

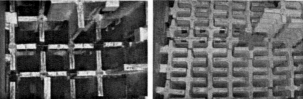

Figure 5 : Instrumentation of the checkers with thermocouples:
for square Type 3 channels (*left*) and for rectangle Type 8 channels (*right*)

The resulting exchanged power profiles from the checker instrumentation are shown in figure 6. They are taken at the same time as the heat balance reported in table 3. They strongly confirm the trend observed during the laboratory tests showing a significant increase of heat exchange intensity especially in the Type 8 section.

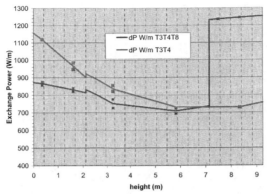

Figure 6 : Exchange power profile in the 2 instrumented chambers standard chamber (*red*) and innovative packing with Type 8 (*green*)

The industrial measurements are in good accordance with the previous modeling predictions. As a confirmation, the simplified model presented above has been applied to the actual working conditions of the industrial furnace. The results have been verified against the measurements and shows very good agreement both in terms of checker temperature profile (Figure 7) and gas temperature. The air preheating temperature values are indeed calculated within 1% spread from the actual measurements.

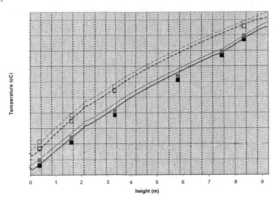

Figure 7 : Comparison of the checker temperature between the regenerator model (*lines*) and the thermocouple measurements (*dots*) in the 2 instrumented chambers: standard (*dash line*) and Type 8 packing (*full line*) at each inversion period (*air in blue, waste gas in red*)

Discussion: Smaller Flue at Risk?

Type 8 cruciform fulfills the expectations of higher thermal performance for a glass furnace regenerator. Furthermore, the Type 8 checker is designed to be placed in the top courses of the regenerator packing. Type 8 is therefore available in ER5312RX material that is suitable for the top course of glass furnace regenerator.

Figure 8 : examples of rows of ER5312 Type 8 mounted over rows of Type 4

While the top of the packing is generally not an area of heavy clogging but recognizing that the size of the Type 8 channel represents approx 40% of the corresponding Type 3 channel, one could argue that the smaller flue size would likely favor more rapid clogging of the packing than regular cruciform channels. Exhaust gases from glass furnaces are indeed prone to carry particles that could deposit on the checkers or to transport chemical species that could condensate in the lower part of the regenerator (e.g. alkali sulfates). The first industrial experience may be too recent to firmly conclude but the issue of the plugging of the channels by carry over has been investigated in another test on an end fired furnace. A patch of Type 8 channels (4x12 channels / located transversally at 1/3 of the length of the chamber near the target wall) has been installed in the top course of a Type 4 regenerator chamber. The furnace has been running for 5 years and observations did not reveal any difference in erosion or damage between the standard flues (Type 4) and Type 8 flues. No plugging of the top courses of the packing was observed.

Concerning the risk of condensation on Type 8 cruciform checkers and subsequent risk of plugging, it can be theoretically reduced by making sure that Type 8 is outside of the condensation zone. For sodalime silica glass operating under normal conditions, the last bottom course of Type 8 should be located above the 1150°C isotherm and not below where the sodium sulfates start condensing (800 – 1150°C).

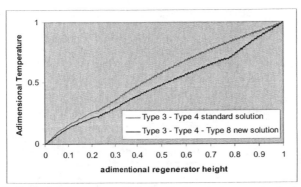

Figure 9 : Typical temperature profile in the checkers between a standard solution
and a solution with Type 8 substitution in the top courses

Numerical modeling helps the designer of the regenerator to choose the right amount of Type 8 for his furnace. In typical furnace, the change from a Type 3 – Type 4 configuration to a Type 8 configuration leads to a modification of the temperature profile in the checkers. Such a change is illustrated in figure 8. Past experience in numerical modeling based on industrial conditions shows that between 15% and 25% of the height of the regenerator can be substituted by Type 8 and having the smaller flues working at temperature above the condensation range.

Figure 10 : View of the top course of a chamber with Type 8 rectangular channels
during the construction (left) and while the furnace is in production (right)

NEW REFRACTORY SOLUTION TO SECURE AND INCREASE LIFETIME

Considering the more and more drastic environmental norms in glass industry, glassmakers tends to reduce NOx & SOx emissions by using some techniques (such as primary settings) that may induce a more or less reducing atmosphere (meaning CO in excess) in the glass furnaces, especially at the top of regenerative chambers.

Our experience (measurements made on furnaces, used products analysis, laboratory tests, and thermodynamic simulation) allowed us to better understand where and what types of species may condense within the checkers as the waste gases cool down, depending on the atmosphere chemistry. Then, if under oxidizing atmosphere the not very aggressive Na_2SO_4 (l) is able to condense, this is no more the case under reducing atmosphere. Indeed, when CO is in excess, SO_2 can be reduced to form H_2S, then NaOH(g) condenses under Na_2CO_3 (l) and above all NaOH (l) form, this last one being strongly aggressive, moreover at a temperature different from the classical sodium sulfate condensation temperature (lower down in the checkers). In fact, the aggressivity of the atmosphere can be described by the index CO/SO2*NaOH(g). This is illustrated by figures 1 and 2.

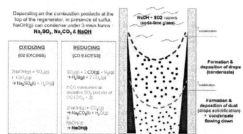

Figure 11 : Condensation mechanisms in regenerators in oxidizing or reducing atmosphere

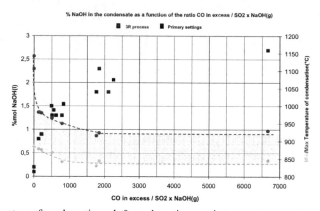

Figure 12 : nature of condensation salt & condensation area in temperature vs atmosphere

Then, this type of reducing atmosphere decreases the lifetime of checkers, whether made with electrofused or sintered products. When a checker made of electrofused cruciform products can resist above 15 years in oxidizing atmosphere, a strong degradation can be observed after only 2 years in highly reducing and alkaline atmosphere (see picture 1).

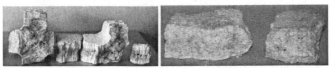

Picture 1 : AZS electrofused product and MgO-Zircon chimney block, 4 years old with 3 years under 3R process

As the corrosion resistance of the current checkers products, located in the condensation area in such application, is extremely limited, we developed on one hand some specific laboratory tests well describing this application, and on the other hand a new product highly resistant for this one.

Lab tests description

- The first test consists in soaking some refractory samples (dimensions ~15x15x80mm) in crucibles containing various alkaline species, as 100% NaOH, at a high temperature (950°C), until degradation. It is named "soak test".

- The second one, named "Channel furnace test", aims to reproduce the atmosphere encountered in a channel of checker (see picture 2). The furnace, with a channel shape, is equipped with a burner working under oxidizing or reducing atmosphere, in the flame of which we inject continuously an alkaline solution (diluted Na_2SO_4, NaOH or Na_2CO_3). On the other side, a chimney extracts the waste gases. The thermal gradient used in general is from 1300°C (burner side) to 700°C (chimney side), so well covers the condensation area. The duration of test depends on the aggressivity of the atmosphere and the shape of samples (small samples or whole pieces). We tested from 24h to 1month duration.

30x80x15mm samples   Wings of cruciform pieces

Picture 2 : channel furnace test

Description of the new ER55XX product and its behavior in lab tests

This electrofused and cruciform shaped product, made of 100% spinel $MgAl_2O_4$ (see chemistry in table 1), doesn't contain any inter crystalline phase (no silica, soda, lime…) what is in general the weak phase for corrosion resistance. Its porosity is high (20%) and fine, well distributed in the wings (Bulk density: 2.8, True density : 3.6). This is shown by picture 4.

Picture 3 : phase diagram & cruciform product (entire / sawed by the middle / piece of wing)

% in weight	Al₂O₃	ZrO₂	SiO₂	MgO	Na2O
AZS	*51.2*	*32.5*	*15.0*	-	*1.3*
β''' alumina	*97.5*	-	*0.5*	*7.5*	*4.5*
Spinel	72.6	-	0.2	27	0.2

Table 4 : chemical compositions

The product shows a columnar solidification, with a high size of crystals strongly imbricate, which is an advantage in case of chemical attack (see picture 5).

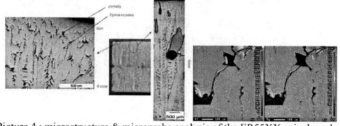

Picture 4 : microstructure & microprobe analysis of the ER55XX spinel product

This product is not only resistant against liquid NaOH (see results of soak test on picture 6 & channel furnace test results on picture 7), but also towards NaOH vapour (see picture 8) or also Na₂SO₄ liquid (see soak test results on picture 9). That means that this product can run under reducing or oxidizing atmosphere, in condensation area or even above. Its very strong resistance against chemical attack is due to the high resistance of the spinel crystal itself, emphasized by the particular columnar microstructure which confers a strong cohesion to the product.

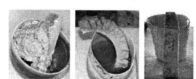

Picture 5 : Electrofused AZS after 20', β''' alumina after 20' and spinel after 8h in pure NaOH(l) at 950°C

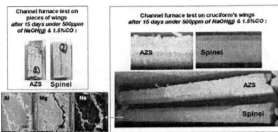

Picture 6 : channel furnace tests on AZS & spinel samples under high alkaline & reducing conditions

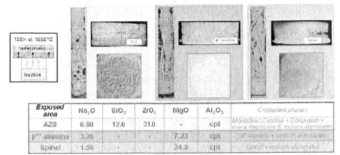

Exposed area	Na₂O	SiO₂	ZrO₂	MgO	Al₂O₃	Crystalline phases
AZS	6.80	12.6	31.0	-	cpt	Monoclinic Zirconia + Corundum + many Nepheline & sodium aluminates
β''' alumina	3.26	-	-	7.23	cpt	β''' alumina + sodium aluminate
Spinel	1.56	-	-	24.8	cpt	Spinel + sodium aluminate

Picture 7 : AZS, β''' alumina and spinel from vapour phase corrosion test

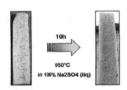

Picture 8 : ER55XX spinel product after 10h at 950°C in sodium sulfate.

Comparison with chimney blocks products

Among the available sintered materials (chimney blocks, bricks…), even the most resistant products in NaOH(l) are very less resistant than the ER55XX (see soak test results in NaOH on table 1). Indeed, even if some of these materials have their coarse grains highly resistant (Spinel grains, MgO grains), they always have a certain amount of bonding phase strongly weaker, which always remain the preferential path for NaOH to go through the products, inducing their damage.

950°C, 100% NaOH(l)

	Electrofused products			Sintered products				
	AZS	β'" alumina	Spinel	Spinel-MgO	Spinel-MgO	MgO-Zircon	MgO low Si	MgO high Si
Test duration	20'	20'	8h	45'	45'	20'	3h	1h

Table 5 : soak test results (in 100% liquid NaOH)

CONCLUSION

Aware of the challenges faced by Glassmakers, SEFPRO developed a new range of refractory materials for the regenerative chambers. ER55XX materials combined with Type 8 shape allows to decrease energy consumption and reduce NOx and SOx emission. This solution already approved by numerous glassmakers is a significant change towards a more environmental friendly production of glass.

LITERATURE REFERENCES

[i] F. Pomar, A. Pinard, F. Martin and Y. Boussant-Roux, International Glass Journal **104**, p. 20-24 (1999)

[ii] Y. Boussant-Roux, O. Citti, M. Miller and S. Chaudourne, Gastech. Ber. Glass Sci. Technol. **73 No 9**, p. 259-269 (2000)

[iii] H. Hausen, Z. Angew. Math. Mech. **9**, p. 173-200 (1929)

[iv] F.W. Schmidt, A. Willmott, in *Thermal Energy Storage and Regeneration*, edited by Hemisphere (McGraw-Hill, New York, 1981), p. 350

[v] Y. Reboussin, J.F. Fourmigué, Ph. Marty and O. Citti, Applied Thermal Engineering **25**, p. 2299-2320 (2005)

[vi] M. F. Modest, Radiative Heat Transfer, second edition

[vii] Métais B., Eckert E.R.G., Forced, mixed and free convection regimes, Trans. ASME – J. of Heat Transfer, vol. 86 (2), pp. 295-296, (1964)

[viii] D. Lagarenne, Thèse INSA Lyon, Récupération d'énergie par les régénérateurs de chaleur des fours de verrerie (1990)

BONDED SOLUTIONS FOR THE CONTAINER MARKET

Thierry Azencot, Savoie Réfractaires

Michele Blackburn, Corhart Refractories

ABSTRACT
Sintered refractories provide an optimum solution for container glass furnaces. They combine long furnace life, low levels of defects with great economic value. This paper aims to detail bonded refractory solution for throat, forehearth and expandable applications.

INTRODUCTION
On the average, the furnace accounts for only 20% of the total investment in the container glass manufacturing process. However, the entire facility's success depends on optimum furnace operation. If the furnace were to fail, or to affect the quality of the glass being produced, there will be serious consequences down the line, affecting production and resulting in increased rejects.

In order to maximize earnings, each individual glassmaker must consider the furnace in terms of both the market and the competition, keeping in mind his own specific requirements such as Furnace Life Time and Glass Quality.

FURNACE LIFE TIME

	Average for modern furnaces*	Trend for the most efficient furnaces**
Campaign length	6 to 8 years	up to 10 years and more
Average specific daily pull	2.8 T/m2	up to 3.5 T/m2
Cumulative specific pull	7000 T/m2	up to 10000 T/m2 and more

Glass production during a given campaign is expressed in terms of cumulative specific pull per square meter of melting area.

Cumulative specific pull is a fundamental criterion for measuring the cost effectiveness of a glass furnace

* Average computed from more than 500 furnaces worldwide
** Average computed from several furnaces equipped with high quality products.

Over a campaign lasting several years, furnaces are subject to a harsh environment and extreme operating conditions. Throughout a given campaign, many hundreds of thousands of tons of molten glass flow through a furnace throat (1170 000 T for a 100 m2 furnace with a specific pull of 4 T/m2 during an 8-year campaign).

The throat is the nerve center of any furnace, and its proper working order depends on precise geometric configurations and dimensions which should last throughout the campaign. Ultimately, these criteria should lead to less wear and tear.

In the past, the throat has often been a cause of shutdown for container glass furnaces. The progressive introduction over the years of high quality 41% zirconia AZS materials and chrome-containing refractories has solved this problem. These solutions have been thoroughly tested on many thousands of furnaces. Because of its excellent corrosion performances, High chrome material are now becoming a « must » for the throat when life time performance is expected to reach ~8000 T/m² and/or 8 years campaign.

Rejects such as green streaks can usually be linked to cracks, in which stagnant glass gets strong Cr_2O_3 enrichment, and is then released when some operating parameters change (temperature, glass pull, etc). The use of high chrome sintered material with ZrO_2 content made by vibro-casting or isopressing leads to improved thermal shock resistance which reduces cracks during operation for better glass quality.

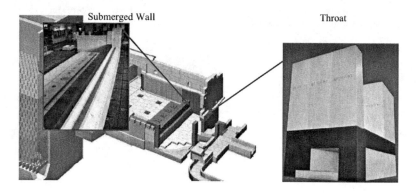

Figure I: High Cr vibro-casted sintered material with ZrO_2 to improve furnace life time

GLASS QUALITY

It is clearly understood that forehearths play a crucial role in the quality of glass produced by the furnace. Any defects generated in the forehearth zones could well be found in the finished product, since the glass at this stage works at relatively low temperatures and has a low capacity to absorb any flaws.

In addition, market demands for quality are constantly on the increase. Defects and flaws which were acceptable in the recent past are now often cause for rejecting the glass. This jeopardizes both yield and profitability of glass production.

Regardless of the design used to channel the glass from the distributor to the machines, and to condition it thermally for forming, the choice of the refractories both in glass contact and in the superstructures is increasingly important.

GLASS CONTACT
One of the types of flaws most often encountered in container glass are flaws known as "cat scratches". Market sensitivity to this flaw is constantly growing, and the flaw affects not only white and flint glass, but also colored glass.

These glassy defects consist of strips of very thin parallel lines -cords (5 to 30 μm) located on the outside surface of the pieces.

 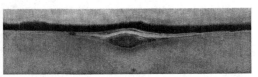

Figure II: Microphotographs of Cat Scratches

Based on several decades of experience in expert appraisals of glass inclusions, glass scientists have developed highly precise typology of these cat scratches. In 85% of the cases, they are characterized by the simultaneous presence of significant amounts of zirconia and alumina. Research into the ZrO_2/Al_2O_3 ratio, and the presence of certain elements which clearly show a fusion origin (V_2O_5 for example in the case of fuel-heated furnaces) have helped identify the origin, which, in 80 to 90% of the cases, is the end zone of the furnace: distributor, forehearths and spout area.
This understanding has led many glassmakers to reduce the use of Zircon-Mullite type materials as much as possible, and generally speaking to reduce the use of any material with significant amounts of zirconia, in their forehearths.

THE SINTERED ALUMINA SOLUTION
Up until the early 1990's, the only option was the use of fused-cast Aluminous (Jargal M type) pieces. This solution is technically sound but a costly solution for container glass applications.
Development of highly pure sintered aluminous product appeared quite attractive, with the objective of combining the advantages of the aluminous products in terms of glass quality with the competitiveness and format production flexibility of sintered products.

Type of Product		Zircon-Mullite	Fused-cast Alumina	Vibro-cast Sintered Alumina
Chemical Analysis	ZrO2	10.5 %	-	-
	Al2O3	75.5 %	95 %	93 %
Physical Characteristics	Density	3.0	3.17	3.10
	Apparent porosity	16 %	3 %	14 %
	Linear expansion at 1000°C	0.7 %	0.75 %	0.8 %
Characteristics in Use	Blistering at 1100°C (0-10)	2	0-1	1-2
	Stoning at 1250°C (0-5)	1-2	0	0-1
	Cat Scratches	+	0	0
	Corrosion resistance (MGR – 72h – 1350°C)	80	100	80

Figure III: Comparison table: characteristics of glass contact products for container glass forehearths
As seen in the table above, sintered alumina made by vibro-casting presents characteristics which are very close to those shown by fused cast alumina.

For forehearth applications, where temperatures commonly occur between 1200° and 1280°C, the corrosion resistance between fused cast and sintered alumina at 1350°C, is actually very small. In addition, the lifetimes seen on installed glass production lines for both solutions are similar. Furthermore, the feasibility of large pieces (channel width: 1524 mm) with reduced wall thickness (150 mm) has been confirmed. This provides the solution for the increasingly common extra-wide lines, which result from the continually increased capacity of the forming machines.

Figure IV: Pre-assembly of a sintered alumina forehearth

Since 1990, more than 700 forehearth lines made with sintered alumina refractories have been installed around the world. Performance in the field has confirmed this product stands out as the optimal solution for container glass applications.

Figure V: Sintered Alumina forehearth after 8 ½ years of clear glass campaigns

The Alumina Concrete Solution
Worn joints between forehearth channel pieces can lead to glass penetration, particularly during long production campaigns. Glass contact with the insulating materials is to be strictly avoided. The solution consists of using a sealing concrete with high alumina content (93%) similar to the content of the sintered alumina glass contact material being used.

		Alumina Concrete	Vibro-cast Sintered Alumina
Chemical Analysis	Al2O3	93%	93%
	SiO2	6%	6%
Physical Characteristics	Density	3.3	3.1
Characteristics in Use	Blistering at 1100°C (0-10)	1-2	1
	Stoning at 1250°C (0-5)	1	0-1
	Corrosion resistance (MGR – 48h – 1300°C)	65	100

Figure VI: Comparison of Alumina Concrete and Sintered Alumina

Alumina concrete blistering rate is particularly low (rating 1 in the standard 1-hour test at 1100°C), comparable to sintered alumina performance.
Application of one safety layer of alumina concrete is the perfect addition to the use of sintered alumina in the forehearth channels themselves.

SUPERSTRUCTURE
Today, changes have occurred in superstructures, tending toward large, monolithic blocks. In addition, the superstructures are tending toward different designs which require complex shapes. These changes are designed for the optimal thermal conditioning of the glass.
A wide range of materials is available to make these large, monolithic pieces, in order to meet the specific needs of many different configurations, such as coloring forehearths.

In Europe, the traditional forehearth superstructures most often use materials with 60% alumina content. However, the changes occurring in the sizes and complexity of forehearth pieces, as described above, and the increased campaign lengths, have necessitated the development of a product showing improved resistance to creep and to corrosion in vapor phase. This has led to the development of higher alumina content material, with typically 67% alumina. Such a material has been successfully developed and launched since 1999, particularly in applications which require extremely large size pieces, such as distributor covers.

		Sintered Alumina 1	Sintered Alumina 2
Chemical Analysis	Al_2O_3	67%	61%
	SiO_2	29%	34%
	Fe_2O_3	0.5%	0.8%
	TiO_2	0.2%	0.6%
Physical Characteristics	Density	2.6	2.5
	Apparent porosity	15%	17%
	Cold crushing strength	100 Mpa	80 Mpa
	Softening under load 28 psi/T: 0.5%	1585°C	1540°C
	Thermal conductivity at 1000°C	2 W.m-1.K-1	2 W.m-1.K-1
	Linear expansion at 1000°C	0.52 %	0.54 %

Figure VII: Comparison of characteristics between 2 sintered alumina materials

EXPENDABLES

Expendables are the very last refractory pieces in contact with the glass, and as such must display the highly specific properties of good corrosion resistance, low blistering rate and low stoning rate.

There is a wide range of materials that can be used. Sintered AZS refractory material is commonly used for Soda-Lime Glass operation. Corrosion performance, so the life time, is known to be a function of the $\%ZrO_2$.

Trends towards light containers have lead to refractory components specially designed to promote optimum gob forming conditions. Besides regulating glass flow, spout assembly refractory must form gobs of consistent weight and shape.

Similarly to the furnace throat, the spout throat is the nerve center of any expandable, and its proper working order depends on precise geometric configurations and dimensions which should last throughout the campaign. Ultimately, these criteria should lead to less wear and tear.

In the past, spout has often been a cause of shutdown for container glass operations. The progressive introduction over the years of high chrome spout insert has solved this problem. These solutions have been thoroughly tested on many thousands of furnaces.

Figure VIII: Spout with High Chrome insert

The use of high chrome sintered material with ZrO_2 content made by vibro-casting leads to the ultimate corrosion and thermal shock resistances. It is a key success factor in promoting optimum gob forming conditions during long time operation for better glass quality.

Life Time index

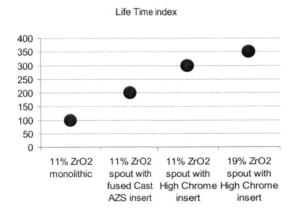

CONCLUSION

Sintered refractory material can provide optimum solutions for the container glass industry. The solutions include optimized chromic oxide refractory for throat and submerged wall, improved monolithic materials for the bottom area, high alumina refractory solutions for glass contact and superstructure application at forehearth, and optimized chromic oxide insert refractory for expendables. The solutions work together to provide an integrated solution to improve glass quality and furnace life for container glass furnaces.

LOW-COST FUSED CAST REFRACTORIES: SOME PECULIARITIES AND CONNECTION WITH GLASS DEFECTS

P. Carlo Ratto
Fused Cast Refractories, Italy

ABSTRACT
Glassmakers are increasingly considering, experimenting and practicing low-cost fused-cast refractory procurement, mostly from Chinese manufacturers. These materials are produced based on locally developed technologies and display peculiar characteristics with practical consequences on the refractory performance, including transfer of defects to the glass. Though there is undisputable financial drive for such a trend, due attention should be paid to minimize the risks/benefits ratio

THE OVERALL SCENARIO

When in 1955 China Building Materials Academy (CBMA) developed the first fused cast mullite block and in 1958 China Shenyang Refractory factory began to manufacture fused cast zirconia mullite blocks utilizing imported Russian technology and equipment, the Chinese refractory industry and specifically the glass refractory segment was an under-developed industry operating in a closed environment, where low quality glass was produced with low quality refractories, under the volume pressure of a rapidly expanding population.

At those times, development of a domestic fused cast industry was, therefore, mostly motivated by the autarchic stance of feeding a growing domestic market with cheap domestic goods, under the strict controlled mid-term planning typical of communist governance.

Things have enormously changed during the second part of the last century. A dramatic acceleration

Fig 1: Old EAF, Russian technology

took place over the last decade, with China joining the WTO, triggering a number of problematical consequences for the western world, emerging countries and China itself. In spite of recurrent and understandable protectionist temptations, it is common feeling that the process of globalization has gone too far to be reverted without worse effects than those arising from the ongoing globalization itself.

The world of refractories, refractory for glass and the glass itself are no exception to this global revolution. China has become the largest overall exporter, heavily leveraging on its peculiar low-cost position, mostly due to a large availability of a crucial raw material at bargain prices: labor.

In coincidence with this globalization process, western economies have undergone apparently endless waves of financial and economical crises, pushing most of industrial systems, including glassmakers, toward evermore strict financial control, thus abandoning the traditional cautious approach to refractory procurement in favor of a more speculative low-cost approach.

Between fifteen and ten years ago, the Chinese fused cast industry, which had meanwhile developed tens of small-medium sized factories spread in several industrial provinces to feed the local glass industry, began to face the opportunity and challenge of feeding a newly available export market, toward a globalized western glass industry eager to access low-cost supplies, and prone to assume unprecedented levels of risk.

Chinese fused cast manufacturers were not at all ready to seize this attractive opportunity of accessing hard currency and safe credit with their delay on sales and marketing capabilities, customer servicing, application engineering, technology and, last but not least, product quality.

Whoever has got direct experience of negotiation with Chinese entities, perfectly knows how much the commercial aspect dominates every facet of a trade. Chinese manufacturers (and vendors) have always been leveraging on price, sometimes even beyond what the real costing position might allow.

Eventually, western glassmakers, accustomed to receiving outstanding levels of technical service and product quality from the few traditional western fused cast manufacturers, had to cope with difficult communication, poor levels of service and exotic standards of quality coming from another technology and different process control practice. Indeed Chinese manufacturers, with their usually common source of domestic technology, had developed processes and products instrumental to their local glass industry, which was meanwhile undergoing progressive changes as well.

THE BLOCKS APPEARANCE ISSUE

It must be noted that the first and foremost marketing impact, when one tries to sell or buy a fused cast refractory block, is its visual appearance, and this was the first concern.

I am not saying this is a critical technical aspect; engineering parameters such as resistance to corrosion, glassy phase exudation, and blistering potential are at the core of the technical evaluation. But it is also obvious how large surface irregularities (bulging, depressions etc.), discoloration, lack of planarity, dimensional and squareness deviations, etc. may

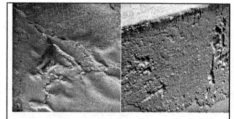

Fig.2: Unacceptable as-cast skin qualities

easily prevent a block from consideration as a candidate for technical evaluation from start.

Shortly after initial approach with western glassmakers, it became evident that the domestic fused-cast surface and geometric standards were unacceptable for any export market. The existing Chinese technology, and particularly the molding process were (and largely still are) inadequate to generate acceptable as-cast surfaces, at least for not-mating faces.

Thus far, the relatively successful technology protection put in practice by the western manufacturers and the poor financial availability and determination of Chinese manufacturers have generally made it impractical to upgrade their molding technology to the level necessary to produce satisfactory as-cast blocks surfaces.

It is general opinion that the kernel of fused-cast technology stays around the smelting/refining process, whilst the most critical process stage is linked to molding and its balancing with subsequent flasking and annealing technology.

The poor control over manufacturing cost factors and the lack of focus on company's financial performances, led these low-cost fused cast manufacturers toward a solution which would have never been conceivable in any western company: instead of improving the molding technology to achieve flatter, cleaner and geometrically acceptable as-cast surfaces, Chinese developers opted for grinding six faces with the intensity necessary to eliminate all the skin defects and restore the geometrical tolerances.

Since grinding fused cast materials, and particularly zirconia/corundum bearing stones, requires considerable amounts of energy and diamond tools, one major trend of western producers of fused-cast has always been to minimize the grinding stock on mating surfaces. Thus the idea of grinding six faces might have been accepted only in the absence of options to improve the molding and lack of pressure on the level of grinding cost. Indeed this happened to be the case, ten years ago, for Chinese manufacturers looking for an export market. And this is still the case, for most of them.

This "innovation" introduced a great improvement in the aesthetic appearance of blocks; not only do ground surfaces appear flat and clean, but also surface cracks and edge cracks disappear or reduce, since the block skin is removed for a variable depth, depending on the skin defectiveness and initial geometric deviations. It is obvious that, without preliminary knowledge of the surface/geometry defectiveness of a specific block, the grinding stock must be planned in excess when designing mold dimensions. Grinding stock that is typically only 2 to 4 mm for mating surfaces for European or US technology normally becomes several millimeters for six faces for an average Chinese block. This means, at least, 5× grinding costs and grinding shop capacity for low-cost vs. western manufacturing!

Fig.3: Six faces ground AZS#41 blocks

This unexpected blocks good looking proved a good seller in front of financially oriented customers, also in the case of companies with advanced capability of evaluating these critical refractories for parameters of real significance such as corrosion resistance, blistering potential and exudation tendency. On the other hand quantification of the latter entails relatively complicated tests with statistically significant number of determinations and correspondingly long evaluation timeframe. Not always are these times compatible with procurement procedures and in no case standard tests are designed to determine the effect of a block scalping.

THE FUSED-CAST BLOCKS SKIN

What can we say about the effect of an AZS block surface scalping, particularly at glass contact and superstructure hot face?

In order to understand the consequence of taking the skin away we must describe what the skin is and, in the first place, mention why a skin exists.

Fused-cast refractories, and particularly AZS fused cast are heterogeneous bodies by nature. They are manufactured by pouring a liquid ceramic of a given composition into a mold. The liquid cools and precipitates grains starting from the liquid-mold interface toward the casting core, where a shrinkage cavity develops because of a different density between the solid (high dens.) and corresponding liquid (low dens.) phases. As a result, deep quantitative differences in chemical, mineralogical and structural parameters are well known within a given block structure, mostly distributed along isothermal surfaces.

Starting from the block surface toward the block core (perpendicular to the isothermal surfaces), in an AZS block there occur different zones and sub-zones:

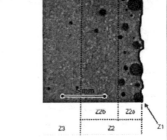

Fig.4: Skin structure subdivision

- Z1: Vitrified layer: this is the result of the first liquid impact onto the silica mold walls. We note that the liquid AZS temperature is about 100°C (212°F) above the fusion temperature of pure silica sand (quartz), which is the mineral component of the mold panel. AZS liquid rapidly quenches once in contact with the mold panel, thus precipitating crystalline grains thanks to the thermal capacity of the cold panel. This thin layer is mostly amorphous and contains some thermally transformed silica grains, soda from the liquid and panel composition (depending on binder type) and a few crystalline components from refractory (corundum and baddeleyite). If the mold quality is decent and the casting technique appropriate, this shiny layer is very thin, a few tens to a hundred microns.

- Z2: The so-called skin: this represents the first liquid frozen in an intense thermal field with precipitation of crystalline species. It is composed of small elongated corundum crystals, oriented normally to the surface and saturated with co-precipitated baddeleyite crystals in a pseudo-eutectic structure. Moving toward the block core, corundum crystals tend to become more isometric and randomly oriented, displaying increasing amounts of inter-crystalline glassy phase patches, i.e. approaching the micro-structure typical of the inner block. Within this skin we observe a further zoning, related to the so-called sub-skin porosity:

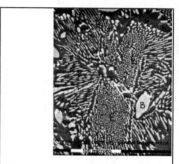

Fig.5: Z2 normal to isothermal surf.

  - Z2a: this is the most external portion and includes meso- and macro-pores (50-1500 microns) which are mostly related to gaseous emission from the mold panel heated by the liquid, and the mold panel gas-permeability. Depending on the panel binding system, gases may include water vapor, carbon oxides, nitrogen and its oxides. The less the panel is permeable, the more these gases penetrate the Z1 and get trapped between the precipitating crystals within Z2. Several efforts are being deployed by western manufacturers to minimize this occurrence and keep this Z2a layer as thin as possible. In general lower amounts of organic panel binder, stronger but highly permeable panels structures, reduced hygroscopicity, shorter storage time and conditioned storage are

Fig.6: Z2b parall. to isothermal surf.

among the ploys to minimize the severity of sub-skin porosity and thickness of the affected layer. In spite of these efforts, sub-skin porosity is still a challenge and whoever is familiar with AZS fused cast blocks inspection knows that sometimes there may exist worrisome patches of macro-porosity in ground mating surfaces (typically sidewalls) as a consequence of opening these bubbles.

  - Z2b: this is the inner skin portion, transitioning into the core structure. It is less affected by sub-skin porosity unless there occurs so-called worm-holing, which is not to be regarded as a physiological scenario. The deeper one gets inside the block, the more there appears inter-crystalline patches of glassy phase, free large baddeleyite crystals, isometric corundum crystals, some of which without co-precipitated zirconia. It is difficult to generalize the thickness of this skin portion, which

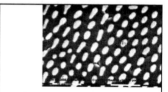

Fig.7: K-B pseudo-eutectic in Z2

depends on a number of factors related to the mold structure, its flasking configuration (therefore the overall thermal conductivity), the block size/shape (i.e. its thermal capacity) and other technological details. Z2b can amount to a significant portion of the skin or be almost nil when, due to a poor molding technology, most of the skin is affected by porosity.

- Z3: this is the block inner portion. It is also heterogeneous, but qualitatively made of pseudo-isometric randomly oriented corundum crystals, some of which with intra-crystalline co-precipitated zirconia, free baddeleyite crystals (and its fern-like structure), glassy phase patches and, getting toward the block core and shrinkage cavity, increasing porosity due to liquid de-gassing and small shrinkage cavity segregations. As a rule of thumb, the more one gets toward the main cavity, the less the structure becomes resistant to glass corrosion.

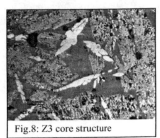

Fig.8: Z3 core structure

In general we can say that Z2 (Z2a+Z2b) extends for millimeters toward the block core, typically seven to twelve millimeters or 1/4" to 1/2". In spite of the presence of pores, Z2a is very resistant to glass corrosion due to its fine grained micro structure, high ratio of pseudo-eutectic corundum/baddeleyite in the form of elongated crystals (oriented normal to the working surface) and lack of glassy phase patches.

Everybody who has performed a post-mortem inspection of an AZS fossil-fired furnace sidewall has noticed that, unexpectedly, vertical mating joints are generally protruding toward the inside of the furnace in a horizontal corrosion profile under the metal line. This points to the occurrence of a sort of a boxing effect (Z2) whereby the tempered skin offers significantly higher resistance to corrosion than do the block internal portions (Z3).

It is out of discussion that, for any given block, the skin structure is the most resistant to corrosion and that, therefore, scalping a block cannot be beneficial nor neutral to corrosion resistance.

But, what about glass defects?

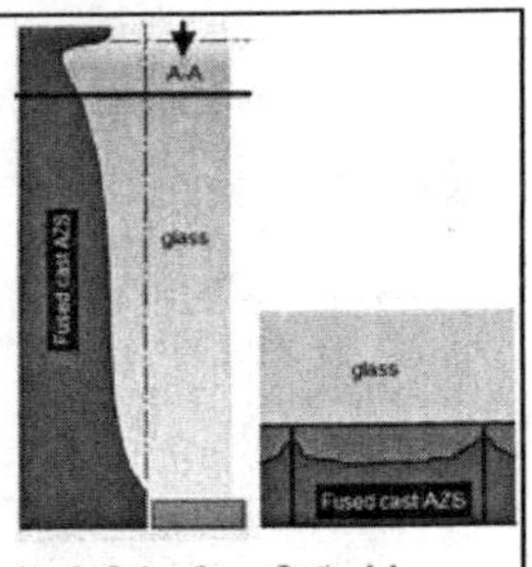

Fig.9: AZS blocks corrosion profiles

EFFECTS ON GLASS QUALITY

Clearly a high corrosion rate implies a high cession of defects (knots, cords, etc.), but there may be further motivations behind the observed high occurrence of zirconia related defects at the early stages of a new campaign, when six-faces ground blocks have been installed in fossil-fired container furnaces:

AT GLASS CONTACT: when the molding technology is not very advanced, Z2a can represent almost the entire Z2 skin layer. Therefore sub-skin porosity can penetrate so deep that, even after scalping away ten or more millimeters, there still occur residual bubble shells.

Please note that even at 90% of a bubble ground off, there still remains a residual shell with pore diameter 60% of the bubble diameter ($D_{pore}= 0.6 \times D_{bubble}$), given that:

$D_{pore}= D_{bubble} \times \sin(\mathrm{acos}(1-2p/D_{bubble}))$

where p is the ground off length of a bubble.

In practice, this means that even a deep scalping in a thick layer of sub-skin porosity may leave a ground surface covered by a large density of relatively large shells.

So, what is the problem with these shells?

Surface grinding is normally performed with a variety of diamond tools working on a horizontal surface with recycled loop water used as a coolant. This water contains a substantial amount of small suspended particles from the ground AZS, contaminated by heavy metals from the abrasive paste, as well as stripped diamonds. In spite of a final rinse with clean water (not always scheduled), particles trapped inside the open shells can hardly be removed, unless it is performed some sort of deep high-pressure water cleaning, which is not at all the case to my knowledge. Besides, concerning the aesthetic performance of the blocks, this layer of adsorbed grinding fines is welcomed by those who practice six-face grinding, since it covers up evidence of fine cracks and the porosity itself, conferring the well known plaster-like surface appearance.

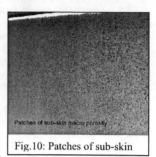

Fig.10: Patches of sub-skin

When all this applies to the glass contact, there is a burst of initial zircon defects, due to the extremely high AZS specific surface exposed to the glass, coming from these fines. The resulting knots and zircon cords can affect initially the glass quality for a longer than usual portion of the campaign. Heavy metal and carbonaceous contamination is also a potential source of blistering.

AT SUPERSTRUCTURE LEVEL: the skin structure, with elongated and oriented crystals, densely packed with its relatively thin layer of inter-crystalline glassy phase, behaves as a barrier against the flow of exudate from the inner portion of the hot face. This skin will not inhibit the extrusion of the liquid phase which indeed appears as a shiny wet surface layer, but will slow down its rate of expulsion and, while acting as a fine mesh filter, limit its content of crystalline species, noticeably baddeleyite grains.

Conversely, the absence or thinning of this skin filter and the surface contamination (see above for the glass contact) will promote a more massive initial exudation and its contamination with grinding fines, particularly rich of zirconia. Exudate knots, polluted by primary zirconia crystals and gas bubbles will contaminate the glass with material of variable average density, capable of affecting the throat flow and eventually getting into the refined glass in the form of cords.

CONCLUSIONS AND PERSPECTIVES

In conclusion, there is clear evidence that scalping working surfaces of AZS fused cast blocks, as it is done by low-cost manufacturers (as a measure to compensate unacceptable quality of as-cast surfaces), is detrimental to glass corrosion resistance and a potential source of glass defects coming from glass-refractory interaction and superstructure exudation. This fact is strongly supported by practical real-world observations.

The entity of these occurrences is variable, depending on the depth of surface scalping and the ground block skin quality.

A thorough understanding of the specific supplier's capacity and the implementation of some special precaution can minimize the risk associated with utilizing six-faces-ground blocks.

However, it is highly desirable that, as soon as possible, these low-cost suppliers will improve their molding technology to the point of making it possible to preserve the as-cast working faces, without unnecessary scalping, which is the way all western manufacturers routinely operate.

This most beneficial improvement is expected to dramatically lower manufacturing costs, release grinding shop bottlenecks and, most importantly, eliminate one principal source of under-performing risks nowadays connected with low-cost AZS procurement.

About the author: Pier Carlo Ratto, with a Science graduation at the Genoa State University in 1977, is an independent consultant for the global glass industry, with a specific focus on fused cast refractory applications, specification and technology. Prior to this, until March 2009, he was a Technology and Technical Marketing Consultant in the area of Fused Cast Refractories at RHI-AG. Prior to this, Dr. Ratto was the Monofrax Business Development Manager within Vesuvius Group for more than 12 years. He has been involved in this area for over 40 years developing experiences in research and development, manufacturing and business development. He has been working for most of the major western players in the field of fused cast refractories for the glass industry.

MULLITE, SPINEL & CALCIUM ALUMINATE – LEADING THE WAY FOR LONG CAMPAIGN/ENERGY EFFICIENT MODERN GLASS FURNACE CONSTRUCTION

Chris Windle, Trevor Wilson, and Rhiannon Webster
DSF Refractories & Minerals Limited
Friden, Buxton, United Kingdom

ABSTRACT

Mullite refractories have obtained a dominant position as the primary choice for regenerator superstructures.long campaigns (16+ years) coupled with low thermal conductivity, chemical stability and specifically thermal shock resistance are attributes that afford utilisation of mullite throughout the regenerator superstructure, including the rider arch, spanner and adaptor courses and increasingly the chimney style packing above this.

Mullite is a generic term covering a variety of compositions; intrinsic requirements are outlined and contrasted with materials that although mullite containing would not cope with potential temperature deviations (+1600 C) encountered in the regenerator. Spinel based refractory compositions offer a wealth of potential applications for glass manufacturing. With the recent expansion of oxy-fuel melters, specifically solar low iron glass; spinel application has increased significantly, now operating in crown and superstructure of melters, producing up to 650 tonnes per day.

Thermo-mechanical and chemical properties will be compared with competing refractory products and design criteria outlined; specifically buffer courses. A habitual problem with float glass production has been the formation of nepheline on the surface of tin bath blocks with the potential of surface peeling, peel attachment to the underside of the glass ribbon and ultimately very costly downtime to re-establish ribbon integrity. Calcium aluminate has recently been introduced into this application and whilst the formation of nepheline is avoided, these compositions must be carefully developed to ensure that hydraulic calcium aluminate phases are not present; otherwise formation of hydrates can disrupt the refractory structure during storage or de-gas in application. A moisture stable calcium aluminate product is described.

1. INTRODUCTION

Mullite based refractory compositions for regenerator construction has been established practice in the container industry for decades. Recently the attributes of these materials have been realised for float regenerators and therefore the tonnage of mullite supply is undergoing a "renaissance." In addition to this the traditional key roles played by mullite in the rider arch, rider/spanner tiles and adaptor courses are being augmented by mullite chimney blocks in the lower packing.

Mullite derived from andalusite can be perceived as exhibiting a dual nature; the mullite crystal exoskeleton provides excellent load bearing and creep resistance, whilst the silica rich crystal interior reacts with the alkali laden atmosphere of the regenerator to form surface "protective" layers.

The formation of these layers is crucial to the longevity of the lining; when fused mullite compositions are utilised for the most arduous applications; i.e port neck/entrance arch, target or division walls, due diligence must be paid to the refractory mineralogy to ensure that alkali attack yields the desired glassy interaction layer "toughened" by corundum needles [1]. It is essential that the

121

corundum phase is derived from the alkali induced breakdown of the mullite structure and that it is not inherent to the composition; specifically it should not be susceptible to ß-alumina interactions which could result in fixed expansion with consequent spalling deterioration of the structure.

Mullite is a truly generic term encapsulating a wide range of alumina-silicate refractories derived from many raw materials; for regenerator use many of these compositions are unsuitable and therefore it is important to define the key differentiation properties; heterogeneous phase assemblages promote heterogeneous reactions and therefore stable protective surface coatings may not form coherently.

Thermo-mechanical properties can also be affected particularly if the mullite phase is limited in extent and this can be illustrated by anomalous creep performance when compared to chemical analysis.

Spinel based on equi-molar $MgO.Al_2O_3$ provides many opportunities for furnace construction. Bonded refractories based on fused spinel aggregate have been successfully utilised for both oxy-fuel and air-fuel superstructure with the longest campaign to date of 13 years.

Variants have been developed to provide complimentary spinel items such a spyhole, camera and burner blocks in conjunction with materials that can be hot inserted.

The fortunes of spinel application is inextricably linked with the growth of oxy-fuel firing; it provides phenomenal resistance to alkali in conjunction with a high melting point (2135°C) and specifically for crown applications, thermo-mechanical properties (creep) at least equivalent to type A silica and αß fused cast.

The recent growth in low iron solar glass has spurned further growth in oxy-fuel firing; spinel has been chosen for both superstructure and crown construction with spans reaching nearly 12m and tonnages up to 650 per day.

Spinel can be termed in refractory nomenclature as a basic-amphoteric material and therefore must be buffered against acidic materials such as silica or silica rich exudates from fused cast products. It is also attacked by batch derived silica and therefore use is not recommended for backwalls or oxy-gas flue systems.

With judicious design and buffer courses, spinel can be applied throughout the majority of the melter superstructure and combined with products which potentially would degrade it by interaction.

Calcium aluminate materials are recent additions to the furnace designers' portfolio; although hypothetical suggestions were posed many years ago, it is only with the introduction of Hibonite raw material (calcium hexaluminate) that compositions have moved from the theoretical to the real.

The target application for these materials is tin bath bottom blocks. Since the float glass process was established in the late 1950s the bath bottom blocks have been a source of process problems. Most of these; "tadpole" defects (albite formation), 7" splitting (unresolved compressive strain) and "open bubble" defects from H_2 thermal transpiration have been solved. One potential issue remains however; traditional alumina-silicate based products are intrinsically susceptible to Na^+ attack which diffuses through the tin from the glass ribbon.

This results in the formation of nepheline on the tin bath block surface which in turn creates crystallographic expansion; this reaction layer can subsequently detach from the body of the block and known as a peel become attached to the underside of the glass ribbon.

If this remains undetected then the ribbon can shatter in the lehr and re-establishment of the process is very costly in lost production time.

Calcium aluminate offers a material with $Na^+/Na\,O$ resistance with a thermal conductivity similar to alumina-silicate compositions which is not possible with spinel.

An intrinsic property of most calcium aluminates is hydration, there is only one phase (calcium hexaluminate, CA_6) that is none hydraulic and therefore refractory compositions should be based on this, otherwise uncontrolled hydration may occur during product transportation or storage.

Although calcium aluminate tin bath block installations are relatively recent, the potential of this material has been realized for the tin bath roof construction which also suffers alkali attack and the potential of nepheline formation.

2. MULLITE REGENERATOR REFRACTORIES

Mullite is an ideal material for the superstructure of regenerators; insulating, creep resistant, low and linear thermal expansion; all attributes which ensure a stable, tight structure which excludes the ingress of extraneous air thus ensuring efficient regenerator operation.

Table 1 below shows the typical physical, chemical & thermal properties of regenerator mullites in conjunction with the zone of application. A previous paper [1] compared in depth the attributes of mullite with other materials available for regenerator construction, this treatise considers the range of mullites available in the generic group and the relative suitability for regenerators.

Table.1

% Mullite	79	81	90	97	Fused Mullite 99	Mullite (A)	Fused Mullite (B)
Bulk Density (g/cc)	2.48	2.50	2.55	2.52	2.60		2.68
Apparent Porosity (%)	13.5	13.4	14.1	18.5	16.2		16
CCS (MPa)	86	80	75	90	83		57
Al_2O_3	55.2	57.7	63.0	70.4	74.6	63-64	76.9
SiO_2	42.2	39.7	35.2	27.7	24.6		21.7
TiO_2	0.35	0.39	0.24	0.27	0.05	1.35	0.2
$CaO + MgO$	0.34	0.31	0.21	0.28	0.13		0.19
$Na_2O + K_2O$	0.65	0.57	0.36	0.48	0.23		0.44
Mineralogy	Mullite Amorphous	Mullite Amorphous	Mullite Amorphous	Mullite Amorphous	Mullite (~100%)	Mullite Corundum Andalusite	Mullite (48%) Corundum(47%) Andalusite
Creep, %/hr @ 1550°C 25th to 50th hour	N/A	N/A	0.01	0.0027	0.0 (1600°C)	Not known	0.0124
Regenerator Applications	Rider Arch Spanner tile Adaptor course Cruciform Lower sidewalls	Rider Arch Spanner tile Adaptor course Cruciform Midwall	Spanner tile Cruciform Upper sidewalls Crown Mid division wall	Upper walls Target wall Upper division wall Crown	Port neck Port neck entrance arch Target wall Upper division wall	Upper regenerator structure	Upper regenerator structure

PROTECTIVE SEALING LAYERS

Mullite compositions react with the degradents in the regenerator system (chiefly alkalis) to form a series of protective layers which over a long period of time conform to the isothermal section of the Na_2O-Al_2O_3-SiO_2 phase diagram at the relevant temperature.

Above a temperature of 1270°C highly crystalline mullites may be used as mullite will breakdown in the presence of alkali to yield a glassy phase with dispersed corundum needles.

The initial phase assemblage is therefore critical to the establishment of these protective layers; mullite/amorphous phases must coherently resist attack to form a barrier of uniform refractoriness. If however the composition consists of a melange of high refractoriness components set in comparatively low refractoriness matrix; the matrix can be preferentially attacked by degradents without participation from the more refractory species.

The consequence of this is that the matrix is rapidly corroded leaving a skeletal framework of the high refractoriness components.

Mullite (A) composition suffered the aforementioned inability to form a protective layer and consequently suffered accelerated alkali attack.

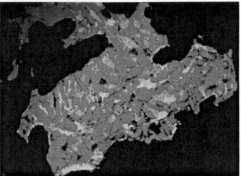

Fig.1 Corundum, skeletal reaction layer; Mullite (A)

MULLITE & CREEP RESISTANCE

Mullite mineralogy/phase is synonymous with resistance to deformation under load; that is creep. Mullite phase developed both on an intragranular and more specifically on an intergranular basis adheres the elements of the structure through a series of interlocking lathe shaped crystals. For the most arduous applications the % of mullite should be as high as possible; the 97% mullite and Fused Mullite products (99% mullite content) exhibit excellent creep resistance (table.1) and this parameter would assure use in the upper division wall; an application in which only a minimal temperature gradient exists hence the key criterion of creep resistance.

Conversely the Fused Mullite (B) product contains less than 50% mullite and therefore although the chemical analysis and physical properties describe a high duty refractory; creep characteristics are not appropriate for upper regenerator use (see fig.2 below).

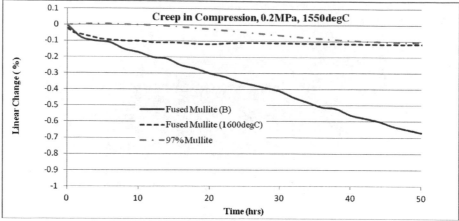

Fig.2

As observed in Fig.2, the Fused Mullite (B) refractory shows continuing deformation over the 50hr time span whereas the high mullite content products have reached a "plateau" of stability in the secondary creep stage.

Creep rate is proportional to the applied stress, however a power relationship exists between creep rate and temperature; H.T.Godard et al [2], stated a 100x increase in creep rate for every 100°C increase in temperature(fused cast); consequently the creep of the Fused Mullite product at 1600°C far exceeds that of the (B) product (@1550°C).

Placing these two compositions in the same generic group is extremely misleading if the application demands a temperature stable product; specifically division walls, and therefore engineers must look beyond simple data sheet criteria and consider mineralogy in conjunction with typical creep characteristics.

Although creep could not be used as a practical control parameter for supply (very costly and time consuming); a small sample set can be taken at random; data provided would ensure that the manufactured batch was characteristic of previous historical supply.

It is also important to consider that occasionally the regenerator superstructure is subjected to temperature excursions (~1600°C) for example when reversal is "hung". These circumstances maybe considered rare in regard to the total furnace campaign length, however it is these circumstances that define the most appropriate product for the application. Clearly Fused Mullite (B) would be severely compromised in such a circumstance.

3. SPINEL MELTER REFRACTORIES

Spinel based on equi-molar $MgO.Al_2O_3$ has numerous potential applications for glass melter construction and these have been outlined previously [3]. Recent expansion in oxy-fuel melters for low iron solar glass has provided further applications for this material; utilisation spreading beyond superstructure to the crown.

The relative merits of spinel as a crown product in oxy-fuel conditions are outlined below in table 2, comparisons are made with Type A silica and Fused Cast αß-alumina.

Property	Type A Silica	Fused Cast αß-alumina	Fused Cast ß-alumina	Spinel
Alkali resistance	Poor	Good	Good-surface reaction	Excellent-no reaction
Creep resistance	Good	Good	No data	Good
Cost	Low	High	High	Medium
Maximum temperature (degC)	1630	+1630	+1630	+1630
Defect potential	Low	Medium	Medium	Low

Table.2

Relative alkali resistance is shown in fig.3 and transmitted light sections of the reaction face in fig.4. Thermo-mechanical behaviour is described in fig.5; spinel attains stability at 1600°C and near zero creep rate at 1700°C.

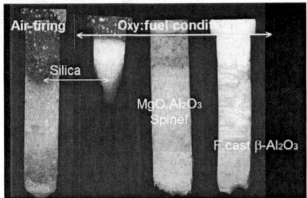

Fig.3   Soda Vapour Test, 1450°C, 7 days   Glasref Consulting

Fig.4 Soda Vapour Test, 1450°C, 7 days   Glasref Consulting

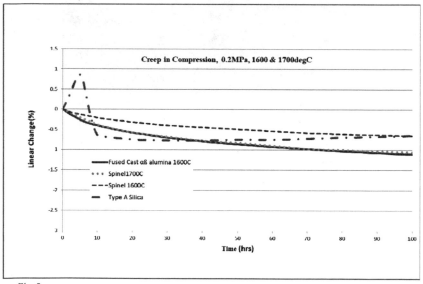

Fig.5

SPINEL & BUFFER COURSES

Spinel can be considered as a semi-basic material in refractory nomenclature or perhaps basic amphoteric. Consequently it must be separated from "acidic" compositions including batch silica and exudates from fused cast products.

Theoretically a silica rich eutectic is formed between spinel and silica at 1355°C; although there are many low melting compositions (1365-1480°C) around the cordierite ($2MgO.2Al_2O_3.5SiO_2$) and sapphirine ($4MgO.5Al_2O_3.2SiO_2$) phase fields.

The recommended buffer layers are as follows:-

Material	Buffer
Silica	Zircon, zirconia tile or >10mm of zircon rammable
LGP AZS	>95% alumina or zircon mortar advised, however will survive without buffer
Mullite	>95% alumina, zircon, zirconia tile or >10mm of zircon rammable
Fused cast alumina (α&ß) or > 95% magnesia	No buffer required

For applications <1300°C buffer layers are not required.

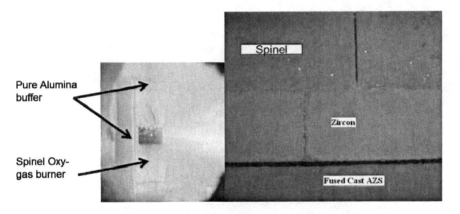

Fig.6   Spinel burner buffered against exudate (LH) and standard Fused Cast AZS

4. CALCIUM ALUMINATE FOR TIN BATH APPLICATIONS (FLOAT GLASS)

Calcium aluminate based materials have been proffered for many years as potential substitutes for alumina-silicate compositions in tin bath block applications. Calcium aluminate is intrinsically alkali resistant and should not undergo the deleterious reaction with Na/O to form nepheline; a process which is accompanied by volume expansion.

Since the development of the float process; the quality of the glass and production efficiency has been inextricably linked (in part) with the performance of the tin bath block. Albite (tadpoles), 7" splitting and open bubble defects are issues that all originate from the bath block. Over time the majority of problems have been solved, however one remains; the formation of nepheline.

Nepheline forms as the result of the diffusion of Na^+ ions from the glass ribbon through the tin; these become oxidised and react with the block to form nepheline. This process results in a volume change in the region of 20-30% which creates stress between converted and unconverted layers.

Large "flakes" of nepheline containing material (Fig.7) can then shear from the block surface and rise through the molten tin to become adhered to the bottom of the ribbon. If these flakes are not detected in time, the ribbon can shatter in the Lehr with subsequent expensive downtime.

Fig.7 Section of a Nepheline flake

Tin bath block compositions have evolved to provide low permeability (lower reaction capability), low reducible oxide (reducing O^{2-} availability) and calcia rich glassy phases to inhibit nepheline formation and indeed the kinetics of the nepheline reaction can be slowed beyond the bath operational campaign.

Thermodynamics dictate however that at some point, nepheline will be the stable reaction phase on the bath block surface.

Consequently calcium aluminate compositions have been developed; these naturally low permeability materials have to be engineered to maintain this feature whilst exhibiting low H_2 diffusivity; a parameter that is inversely proportional to the former.

In addition, whilst the adoption of calcium aluminate provides the much desired avoidance of nepheline formation; compositions are prone to post manufacture hydration which is a corollary of the calcium aluminate (CA) system.

Only one phase is non-hydraulic (hydration stable) and therefore considered a truly "safe" option for manufacture, transport and storage of tin bath blocks; Calcium Hexaluminate (Hibonite).

Table 3 below shows the properties of a calcium hexaluminate material in comparison with a widely used alumina-silicate tin bath product; fig.8 shows bath blocks in production.

Property	Calcium Hexaluminate	Alumino-silicate
Al$_2$O$_3$	81.9	40.9
SiO$_2$	5.75	54.4
CaO	10.5	1.4
Bulk Density (g/cc)	2.27	2.21
Thermal expansion (%) 20-1200°C	1.01	0.70
Thermal conductivity (W/mk) @ 1000°C	1.16	1.39
Hydrogen Diffusivity (mmH$_2$O)	<150	<150
Air Permeability (nPerm)	0.08	0.2

Table.3

Fig.8

Recently the attributes of the calcium aluminate materials have been recognised for other parts of the tin bath construction; the roof. Traditionally lined with interlocked suspended tiles in sillimanite compositions, the roof construction has sporadically also suffered from alkali attack and nepheline formation and flaking.

Vapour attack tests (fig.9 below) emulating the tin bath roof application; show on sectioned test samples that the calcium hexaluminate material has little or no reaction. The fused mullite and sillimanite materials have reacted with the formation of a surface layer (top of section).

Calcium Hexaluminate   Fused Mullite   Sillimanite

Fig.9 Vapour attack tests,Na₂O, 1100°C, 7 days

This behaviour is clearly indicating that the calcium hexaluminate material is intrinsically stable in alkali environments and therefore a potential replacement for sillimanite compositions to combat nepheline induced flaking.

5. CONCLUSIONS

Mullite based materials are proving increasingly popular for regenerator construction; not all mullites however are suitable, and the term "mullite" is generically far reaching. Heterogeneous phase compositions should be avoided as unpredictable/non passive layers may form on the refractory surface with resultant active corrosion.

Data sheets provide chemical analysis, however this is not necessarily correlated with phase assemblage specifically mullite content, which is key to creep resistance and these characteristics should be known, if the proposed mullite is intended for arduous applications such as the regenerator upper structure, crown or division walls.

Spinel utilisation in oxy-fuel melters is increasing based on it's thermodynamic and thermo-mechanical stability in conjunction with a low defect potential. Initially forming the downstream melter superstructure in which the longest campaign is now 13 years; spinel has now been used in the crown (11.7m) of oxy-fuel melters dedicated to solar glass manufacture.

Calcium Hexaluminate compositions are relatively recent products, however they offer the opportunity to negate an old problem of nepheline formation on alumina-silicate tin bath blocks. Recognising this facet of the material, consideration is being made for the application of this material for the tin bath roof.

REFERENCES

[1] Windle & Wilson, "Mullite Regenerators- An Optimum solution", 70th Conference on Glass Problems (2009), 203-211

[2] Godard,Kotacska,Wosinski,Winder,Gupta,Selkregg,Gould, "Refractory Corrosion Behaviour Under Air Fuel and Oxy-Fuel Environments", Ceram.Eng.Sci.Proc,18[1] (1997) 180-207

[3] Windle, "Spinel Refractories & Glass Melting", 65th Conference on Glass Problems (2004) 91-105

ACKNOWLEDGMENTS

Fig.1   Ceram Research, UK
Fig.3&4 Glasref Consulting, Mr Geoff Evans, UK
Creep Curves; Ceram Research UK, DIFK Germany, Orton US

MoZrO_2 - A NEW MATERIAL FOR GLASS MELTING ELECTRODES and MOLYBDENUM
GLASS TANK REINFORCEMENTS - EXPERIENCES AND INSIGHTS

Mike Ferullo and Rudolf Holzknecht
PLANSEE
Reutte, Austria

The history of glassmaking can be traced back to the Stone Age and has been used already by the Stone Age societies across the globe. Despite this long history the glassmaking processes are still being constantly evaluated and optimised. Various topics dominate current discussions. Key words as higher strength, low weight, less impurities greater cost effectiveness and emission reduction are frequently heard at glass conferences and in discussions between glass producers. Competition from alternative materials such as plastics for bottles is increasing rapidly. Every glass producer is forced to improve product quality, but at the same time to reduce the costs for consumers. Possible ways to increase the cost effectiveness, for example, are to reduce the production reject parts and prolong the campaign length of a glass tank.

In this two papers we will look at MoZrO_2 Glass Melting Electrodes and at Molybdenum Glass Tank Reinforcement

I. MoZrO_2 - A NEW ADVANCED MATERIAL FOR GLASS MELTING

MoZrO_2 has been developed especially to be used for glass melting electrodes. Doping molybdenum with small amounts of ZrO_2 improves its properties considerably. It offers higher corrosion resistance against molten glass, as well as increased high temperature strength. The excellent thermal and electrical conductivity as well as the good machining ability of molybdenum remain unchanged. Therefore MoZrO_2 is the most suitable material for solving corrosion problems in glass melting electrodes.

Superior high temperature strength
The steady state creep rate of MoZrO_2 at 1600 °C / 2912 °F is approximately 5 times lower than of pure molybdenum and the tensile strength and the 0.2 % proof stress are considerably higher.[1] The usage of MoZrO_2 allows the active lengths of an electrode in the melt to be increased and drastically reduces the risk of sagging during operation.

Mechanical properties of Mo and Mo 5 Vol.-% ZrO2				
Creep test at 1600 °C / 2912 °F in vacuum Creep stress of 15MPa		Notch tensile test at 800 °C / 1472 °F in vacuum		
	Creep Rate [s^{-1}]	Tensile Strength [N/mm^2]	0.2-Proof Stress [N/mm^2]	Elongation [%]
Mo	4.4 x 10^{-6}	226	97	16.9
Mo 5% ZrO_2	9,5 x 10^{-7}	335	246	6.3

Mean values of samples taken from the middle section and the edge of a rod, diameter 43 mm.

Excellent corrosion resistance
Compared to pure molybdenum MoZrO_2 shows higher corrosion resistance in white and green glasses as well as in molten Sb_2O_3 and Sb_2O_3-refined glass. In 2001 Martinz / Matej / Leichtfried discovered

that MoZrO$_2$ considerably reduces the corrosion rate in glass melts.[2] In white sulphate-refined container glass the corrosion rate is decreased by approximately 25 %, in green glass by approximately 45 %.

New investigations by Prof. Dr. J. Matej at the Laboratory of Inorganic Materials of the Institute of Inorganic Chemistry ASCR in Prague demonstrate the corrosion behavior of MoZrO$_2$ against antimony.[3] Matej compared pure molybdenum and MoZrO$_2$ in molten antimony at elevated temperatures. In demanding isothermal and thermal cycling tests the diameter reduction of MoZrO$_2$ samples was 43 % lower than of pure molybdenum. Optical and electron microscopy were carried out in order to determine the mechanism of corrosion prevention. The most likely conclusion is that the combination of a fine grain structure and the presence of chemically resistant ZrO$_2$ in the grain boundary damage.

Dimensions of sample after testing in molten antimony

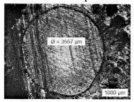  

PLANSEE MoZrO$_2$-doped molybdenum   PLANSEE standard molybdenum   Molybdenum (other supplier)

Comparison of the corrosion behavior of molybdenum materials in molten antimony Test condition: 12 h soaking at 1200 °C / 2192 °F; + 14 h thermal cycling between 880 °C / 1616 °F and 1000 °C / 1832 °F with a half cycle period of 24 min [3]			
	Reduction of sample radius (µm)		Factor
	I	II	average	
Molybdenum (other supplier)	1692	1420	1556	1.8
PLANSEE standard GME molybdenum	1440	---	1440	1.6
PLANSEE ZrO$_2$-doped molybdenum	943	821	882	1

[1] Gohlke et al, 1997, 14th Plansee Seminar Reutte, A
[2] Martinz/Matej/Leichtfried, 2001, 15th Plansee Seminar Reutte, A
[3] Matej et al, 2007, Glass Conference Teplice, CZ

II. MOLYBDENUM GLASS TANK REINFORCEMENT

Great effort has already been made to increase the lifetime of a glass tank. The quality and corrosion resistance of the refractory bricks has been improved year by year. Sintered refractory bricks have been replaced by cast refractories to protect the most heavily worn sections of a furnace in particular. But even the cast quality exhibits corrosion rates that cannot be ignored. The only way to drastically improve the corrosion resistance of glass tank parts that are exposed to heavy wear is to protect them with metal. Only a few metals can withstand the high temperatures required for the production of glass.

Illustration 1 shows a comparison of the corrosion resistance of different metals and AZS material to the most commonly used glass melts. This illustration shows how limited the possibilities are.

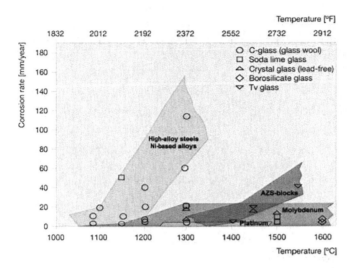

Illustration 1: Corrosion resistance of different metals and AZS material

Iron & iron-based alloys and nickel & nickel-based alloys are not able to fulfil the high requirements of glass production. The melting point of these metals or alloys is much too low and even when used below this temperature, they show high corrosion rates and pollute the glass heavily when immersed in the melt. Two metals have already been proven to exhibit good corrosion resistance without pollution of the glass melt - platinum and molybdenum. The corrosion resistance of platinum is unrivalled, but closely followed by molybdenum. Two main differences exist between these metals:

> the oxidation resistance
> and the price

Platinum is the only metal currently used which can withstand corrosion and oxidation. The oxidation resistance (illustration 2) of molybdenum is poor and therefore it requires special protection from oxidizing atmospheres until it is immersed completely in the glass melt. Nowadays this can be achieved with a coating, called SIBOR®, which will be described in the following paragraph. The big difference between molybdenum and platinum is the price. Platinum is a precious metal and must be priced at a market rate, which is determined day-to-day by the stock market. 10 to 15 grams of platinum cost approximately the same as 1 kg of molybdenum sheet already coated with SIBOR®. The quantity of platinum required is much too high for most uses and it is therefore limited to very special applications such as platinum feeders for special glasses with very high purity levels. Another big disadvantage of Pt is its weak creep resistance. To clad large areas of the tank is due to this weakness impossible while Molybdenum can do this easily due to its high strength and creep resistance.

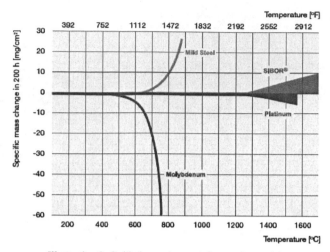

Illustration 2: Oxidation resistance of several materials

SIBOR® coating

As mentioned above and shown in illustration 2, molybdenum has a poor oxidation resistance at temperatures above 600°C (1112°F). In principle, various solutions exist to protect refractory metals like molybdenum against oxidation: alloying, packing with ceramic powders, glassification, cooling, application of protective gases (e.g. hydrogen, argon, ...), cladding with platinum or - last but not least – coating with an impervious layer.

The only coating which can guarantee an "oxidation free" period is the so-called SIBOR® coating. This patented coating consists of silicon and 10% boron by weight and it is applied to sand blasted molybdenum surfaces using a plasma spraying process. The coated molybdenum parts are then annealed to ensure outstanding oxidation resistance. Illustration 3 shows three cross sections of a molybdenum sample with SIBOR® coating after each production step [3a.. SIBOR® coating as sprayed, 3b ... SIBOR® coating after annealing, 3c ... SIBOR® coating after 400 hours in use in air at a temperature of 1450 C (2642°F)].

Illustration 3: Cross sections of molybdenum with SIBOR® coating

The SIBOR® coated molybdenum parts (glass melting electrodes or glass tank reinforcements) can be installed in a cold glass tank before the up-tempering process starts. The parts will remain in tact without any oxidation throughout heat-up, even with a slow up-tempering rate of 5 C to 10°C per hour. It is guaranteed to last as follows: 5000 hours at 1200°C (2192°F), 500 hours at 1450°C (2641°F) and 50 hours at 1600°C (2912°F) after heat up, so at isothermal conditions in the tank. This time/temperature performance enables glass producers to install the molybdenum parts easily and safely in a cold tank. Further advantages of the SIBOR® coating are the properties of the layer. The SIBOR® coating is not as brittle as ceramic coatings like the SiCrFe coating and silicide coatings (e.g. MoSi) and can withstand normal handling during the installation process without chipping. The SIBOR® coating will be dissolved by the glass within a few days. At the beginning bubbling can occur, but it will decrease rapidly after 24 hours. Due to the composition of this coating (Si, B) it will not cause any discoloration or contamination of the glass melt.

Glass tank reinforcements
The lifetime of different glass tanks can vary greatly between a few months and several years. It depends on many factors such as the glass composition and temperature, but also daily production quantities. Opal glass, for example, is a very aggressive glass and a tank campaign lasts only a few months. Glass tanks for container glass (soda-lime glasses) have a service life of up to 10 years (6-8 years on average). The duration of a tank campaign is determined by the rate of wear within the tank and the subsequent failure of important functions.
The performance of the SIBOR® coating mentioned above now makes it possible to use the good corrosion resistance of molybdenum to protect the areas of a glass tank that are exposed to heavy wear in most glass melts (illustration 1). Some of these areas are critical for the lifetime of the glass tank, others are critical for the performance of the tank and the glass quality.

The critical sections are marked in illustration 4, which shows a schematic diagram of a typical glass tank:

 1 throat channel
 2 bubble maker and wall
 3 dog house

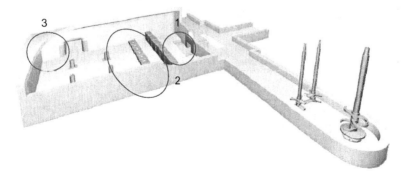

Illustration 4: Schematic diagram of a glass tank showing critical areas

- The "throat channel" – the transition between the melting area and the working end and the feeders and essentially for the glass quality – heavily stressed by corrosion and erosion processes
- The "cross wall" – controls the convection streams in the melting area and the transition time of the glass melt, most important for a good glass quality – heavily stressed by corrosion and erosion processes
- The "bubble maker" – where large, defined bubbles are added to the molten glass to agglomerate the small bubbles – heavily stressed by corrosion and erosion processes
- The "doghouse" – where the raw materials for glass production are introduced into the glass melt – heavily stressed due to the oxygen content of the batch and the formation of foam

The appearance of the wall and the throat channel at the end of the tank campaign as shown in the illustrations 5 and 6 are very familiar to all glass producers.

Illustration 5 and 6: Corroded wall and throat channel at the end of a tank campaign

Cladding with SIBOR coated molybdenum sheet protects these critical areas of the glass tank against wear, maintaining their form and reliability for longer periods. This of course helps to optimize the manufacturing process and glass quality and significantly improves the service life of the glass tank.

Samples for glass tank reinforcements:
Generally speaking, glass tank reinforcements are molybdenum sheets (thickness 6-10 mm, 0.25-0.4"), which are produced in various shapes and designs using bending and machining processes. 100% of the surface is then coated with SIBOR® to achieve complete oxidation resistance of the whole assembly.

Tank components made of molybdenum can easily be fixed to the tank using different methods:

1) Clamping between the refractory bricks
2) Fixing with bolts which are inserted through the sheet into the refractory brick
3) Simply covering of the parts requiring protection such as the wall or the bubble maker

Illustration 7 shows a doghouse reinforcement. Simply formed molybdenum sheets are fixed with pins onto the corner bricks of the doghouse. This helps to prevent corrosion to the bricks as the batch enters and often produces foam in this area.

Illustration 7: Dog house reinforcement

Illustration 8 shows a simple bubble maker cover. The same principal can also be used to protect cross-walls. The U- shaped sheet is simply placed over the refractory bricks and clamped between the bottom bricks. It ensures that the form and function of the bubble maker (or cross-wall) are maintained for a long period.

Illustration 8: Bubble maker / cross-wall protection

The most critical area in a glass tank is of course the throat channel. This section is located between the glass tank itself and the working end and controls transition time and glass flow. The throat channel construction consists of several refractory bricks (two side bricks, one top brick and several bricks for the channel). To guarantee good corrosion protection it is also necessary to cover the joins in the brick. This is only possible if the reinforcement plates used have a greater width and height than the individual bricks and they have to be fixed to the outside of the bricks (cladding). Illustration 9 shows how a throat channel can be effectively protected.

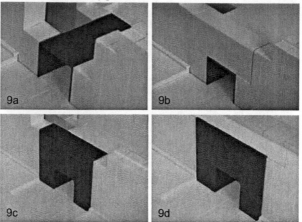

Illustration 9: Schematic installation of a throat channel reinforcement

The throat channel protection assembly consists of two parts: the front plate and a "U"-channel (Illustration 10). This is necessary for ease of handling and installation. The connection between the two parts is specially designed and produced to prevent any penetration of the glass.

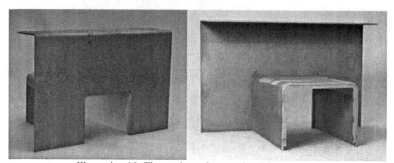

Illustration 10: Throat channel protection - front and back views

Similarly other components such as flow pipes inserted in a cross-wall (used to generate additional convection currents) and delivery pipes can be produced in molybdenum and protected against oxidation. In such cases SIBOR® coating is required on all surfaces.

The design of each glass tank reinforcement is adapted to suit individual customer requirements after discussion between the glass producer and/or the furnace constructor and PLANSEE.
Such parts can be installed easily by either the glass producer, the furnace constructor or by PLANSEE.

Process Control

CHANGING OF GOB TEMPERATURE FROM SPOUT TO BLANK

Gesine Bergmann, Hayo Müller-Simon, Nils-Holger Löber, Kristina Kessler

ABSTRACT
The distribution of glass and the weight of a container depend on the fluid dynamics during forming. The temperature distribution has a strong influence on the glass forming process due to the strong interdependency between temperature and viscosity. Thus, in the first part of this work the measured temperature in the feeder channel is connected with the temperature distribution in the gob by means of a fluid dynamic and thermal model of the spout. This enables to predict the temperature distribution in a gob. In the second part of this paper the working conditions at different IS machines are investigated: i.e. temperature distributions inside the gob, temperatures of the trough, running times and gob lengths at different points of the delivery system. The measured data were processed in a FEM program in order to calculate the temperature distribution of the gob at any point in the delivery. This allows the determination of the heat transfer coefficient between gob and the delivery system by adjustment to the specific conditions.

INTRODUCTION
Technical requirements on container glass ware are of manifold nature. The normal stresses and strains of bottles are based on vertical load, internal pressure and impact; jars additionally have to bear up against thermal shock. Stresses and strains caused by load and the specific wall thicknesses are significantly involved in the limitations of life time of glass containers. So it is necessary to produce bottles and jars with a very uniform distribution of the wall thickness, but not thicker than required. In the practical experience the containers show a distinctive wall thickness distribution in height as well as around the perimeter (Figure 1 – horizontal cut).

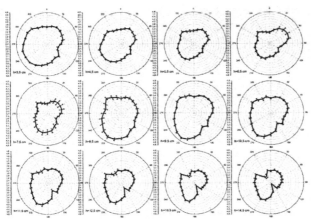

Figure 1: distribution of wall thickness for different heights of the cylindrical part of a bottle

147

Where do the different wall thicknesses come from? To answer this question one has to consider the properties of glass which are responsible for the formability and the manufacturing process as well.

The most important property in this field is the viscosity, because it is the crucial factor for the workability. On one hand the viscosity of a glass is defined certainly by its composition, on the other hand it is depending on the temperature. This correlation is described by the Vogel-Fulcher-Tamman-equation [eq.1]. In case of such a forming process this equation must be extended because of the dependency of the viscosity of the glass melt on the specific time and position inside the glass portion [1]. Therefore, the viscosity or temperature dependence must be examined for every volume element of the glass.

$$\log \eta (T_{\vec{r},t}) = A + \frac{B}{T(\vec{r},t) - T_0}$$

eq. [1]

$$\text{with} \quad A, B, T_0 \quad \text{constants for a given glass composition}$$
$$T \quad \text{temperature}$$
$$\eta \quad \text{dynamic viscosity}$$

If the glass is too cold glass cracks can originate particularly in the finish area, if the glass is too hot it will stick at the mould material and will increase the corrosion rate [2]. A perfect forming of an article needs a certain viscosity window. Based on t he fact that the glass composition of one production should always be the same and should fluctuate only in the limits of the available raw materials, it is worthwhile to have a look at the temperatures of the process, their origin and their development.

Five years ago the HVG started comprehensive investigations of this section of production. In a first joint project with the Fraunhofer-Gesellschaft temperatures at gob perimeters of a double gob system were measured [3]. A second project examined the development of the temperature distribution of the gobs in the delivery. Current studies are concerned with the processes inside the feeder channel and the spout.

The most important result of these projects is the closed mathematical description of the thermal variations of the glass melt inside the spout and the gobs on their way through the delivery system. Our project should contribute to the understanding of the processes in the feeder bowl and the delivery.

FUNDAMENTALS OF WORK

After leaving the melting tank and refiner the glass flows through the feeder channel towards the machines. In this area the glass should reach the temperature required for the forming process. This is simply done by passive radiative cooling on one hand but also active with the input of energy by fossil or electric heating on the other hand. Immediately before entering the spout there is the last continuous control of the glass temperature. This measurement is used as a base for the upstream control process. However, this measurement gives not a picture of the thermal situation of the glass gob before the start of the forming [4].

During the industrial measurements all important parameters are recorded. Depending on the intend of the measurement these include:

- the actual temperature distribution of the glass inside the feeder channel before entry into spout
- feeder tube speed and direction of rotation
- tonnage, gap height and heating of the feeder and the spout
- article weight, machine speed, gob length, gob velocity

- temperature of the gobs, measured with different pyrometers (wavelengths: 1.5 microns, 3.9 microns, 5.14 microns and a 3-fold pyrometer with shorter wavelengths) for different radiative penetration depths.

The positions of the measuring points are shown in the Figure 2.

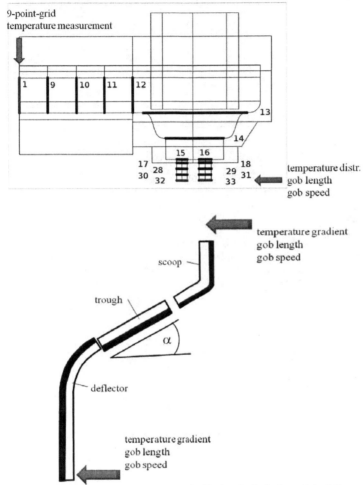

Figure 2: measuring points for the temperature distributions in the feeder and the delivery

The starting point of the model based interpolations is the measurement of the temperature distribution after the shear cut. In the very first step this allows the recalculation of the cross-sectional temperature distribution of the glass strand when leaving the orifice. The interpolations between the temperature measurement inside the feeder channel, the spout and the measuring point after the shear cut are realized with the open source software Elmer, the geometries are captured with the program Gmsh. The interpolations between the gobs temperature distribution at the end of the deflector and the distribution of temperature after the shear cut are done by the free package Calculix.

Fundamentals of the calculations between and inside feeder and spout are the equation of heat transfer and the Navier-Stokes equation. For the calculations of flows in glass melts the model of Newtonian fluids is usually used. The heat transfer equation also allows the interpolations of the temperatures in these areas where the gob is not accessible for pyrometric measurements. The general heat transfer equation needs an additional part for semi-transparent media (eq.2., [5]). The added term describes the radiation at each point of the volume.

$$c_p \cdot \rho \cdot \left(\frac{\partial T}{\partial t} + u\nabla T \right) = \nabla\left(k \cdot \nabla T\right) - \nabla q_r$$

eq. [2]

with

T	temperature
c_p	heat capacity
ρ	density
t	time
u	velocity field
k	heat conductivity (phononic)
q_r	radiation flow vector

Heat transfer is composed of heat conduction and heat radiation. It is widely used to represent the conductivity k as an effective conductivity k_{eff}. That means, that the thermal radiation conductivity is added to the "normal" heat conductivity. The thermal radiation conductivity for glass is frequently determined by the Rosseland approximation [6]. The radiation flow vector is given by the radiative absorbed and emitted flow.

$$c_p \cdot \rho \cdot \frac{\partial T}{\partial t} = k \cdot \nabla^2 T + I_{abs} - I_{emi} = k_{eff} \cdot \nabla^2 T$$

eq. [3]

with

T	temperature
c_p	heat capacity
ρ	density
t	time
I	radiativ flow (absorbed, emitted)
k_{eff}	effec. heat conductivity

The Navier-Stokes-equations describe the flow f.i. in the feeder channel. Additionally the equation of continuity is needed [5].

$$\rho\left(\frac{\partial u}{\partial t} + u\nabla u \right) = -\nabla p + \nabla\left(\eta\left(\nabla u + \nabla u^T\right)\right) + \rho f$$

$$\frac{\partial \rho}{\partial t} + \nabla\left(\rho u\right) = 0$$

eq. [4, 5]

with T temperature
               ρ density
               t time
               u velocity field
               k heat conductivity (phononic)
               p pressure
               η dynamic viscosity
               f volume force

For the solutions of these equations material data are needed. Some of the data such as density, specific heat, thermal conductivity and emissivity were found in the literature. One of the problem is the determination of the radiative penetration depth of the pyrometer because the heat conductivities given by literature have high latitudes even for the same glass colors. Normally the approach of Eddington-Barbier is used for the determination of the radiative penetration depth. The calculations showed however that the temperature profiles obtained in this way differ considerably from the temperature profiles obtained during strong cooling. Because of it was not possible to determine useful results with the spectral data given in literature and using Eddington-Barbier we had to find our own way. A set of material data can be derived if a sufficient number of measurements in different states of the system are available. These data are in good agreement with literature and gave consistently plausible results.

In the delivery system, different directions of motion of the gobs are given. These are crucial for the temperature correlation of the temperature measurements and calculations. Theoretically, the movements may be limited to the free fall and the movement on the inclined plane. Movements of the gobs caused by other impulses (self-rotation, impact at the trough, free flight) were ignored. For this reason there are only two necessary equations:

$$m \bullet \frac{d^2 s}{dt^2} = m \bullet g \qquad und \qquad m \bullet \frac{d^2 s}{dt^2} = m \bullet \sin(\alpha) \bullet g - \mu_R \bullet m \bullet \cos(\alpha) \qquad \text{eq. [6]}$$

with t time
               m mass
               g acceleration of free fall
               s length
               α angle
               μ_R coefficient of friction

MEASUREMENTS

In Figure 3 an example of a temperature distribution on the basis of the 9-point grid measurement is given. Experience has shown that the data recorded by the installed thermocouples show no significant change of the temperature distribution with the time.

As mentioned before the starting point of all the measurements is the determination of the temperature distribution in the gobs after the shear cut. An example of such a measurement is given in Figure 4. The measured temperature differences were determined up to 30 K in the near of the surface (approximately 1.5mm below). Also along the gob center line the differences amounted up to 10 K [3]. The latter can be explained easily. After the extrusion of the gob the already extruded part cools down by loss of radiative heat. The slower the gob is formed, the higher will be the temperature differences along the gob length.

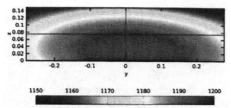

Figure 3: measured temperature distribution of a amber glass feeder (grid)

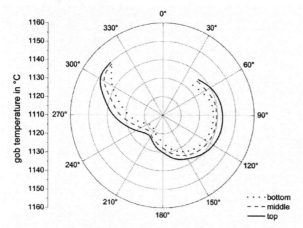

Figure 4: temperatures at the perimeter of a gob, measured with one wavelength

The amount of information increases extremely by using pyrometers with different wave-lengths. A measurement at one machine but on 2 consecutive days is shown in Figure 5. In this figure the temperatures in different depths can be seen as a function of the peripheral angle. It is clearly evident, that the temperature of the gob is not symmetric (zero degree = direction to annealing lehr, 180 degree = direction of furnace). At the surface of the gob the lowest temperatures were detected. The influence on the gob temperature distribution is most pronounced at the surface. A difference in temperature of more than 60 K is possible between the surface and a layer 3mm far away from the surface in amber glass. Depending on the type and color of glass more or less influence on the surface quality of the products can be expected.

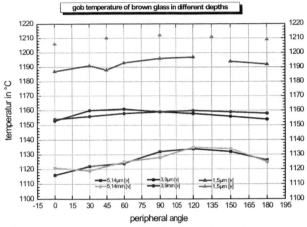

Figure 5: temperature distribution of a amber glass gob after the shear cut

The temperature distribution of the gob after the shear cut will be changed continuously during its course through the delivery system on the following machine layout. The interaction with the environment leads to a release of energy by radiation. In addition there is a cooling of one side of the gobs based on contact with metallic material of the delivery. The temperature distributions of the gobs after the shear cut and at the end of deflector were determined (see Figure 4, Figure 5). Furthermore, the weight of the gobs, the gobs length and velocity were measured. Partly it was possible to detect additional nodes at the start or the end of the trough.

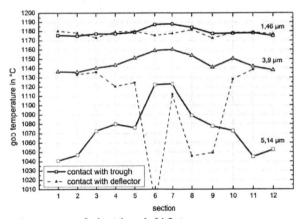

Figure 6: temperature of gobs at the end of deflector

Measurements at the end of the deflector mainly show results comparable to Figure 6. Here one can clearly see that the temperature before the fall into the blank depends on the overall machine layout, that means the length of the troughs in general, but at the same time there is a section specific influence that is more or less distinctive.

An important factor for the formation of the temperature gradient in the gob is the reheating. This was also measured in the delivery, particular at the deflector. The temperatures were measured on the surface of the gobs, which has been in contact with the trough.

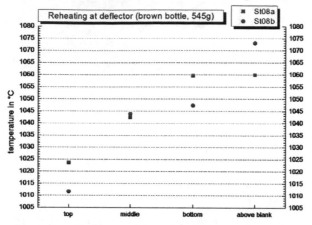

Figure 7: reheating of an amber gob during the transfer of the deflector

The reheating allows an temperature change of 50 to 60 K on the surface while the gob is moving in the last part of the delivery.

In some experiments, the heat transfer coefficient of the delivery has been varied by changing the amount of water-oil-suspension sprayed into the delivery f.e., by cooling a trough with air or by using different coatings. The heat transfer coefficient and the temperature of the delivery have a decisive influence on the surface temperature of the gob.

CALCULATIONS

It is well accepted that the operation procedure inside the spout is a key element for the formation of temperature distributions in the gobs. Already in the 70's extensive studies lead-managed by Greschat were made to present the temperature and velocity distributions in such kind of feeder channel [6]. For these investigations thermocouples were installed at many positions of the feeder channel. Based on these measurements, isotherms were developed over the channel cross-section (Figure 8). These studies showed the shift of the temperature ranges and the development of the flow states.

In recent years the furnace manufacturers therefore built in additional threefold thermocouples at the end of the feeder channel to be able to detect glass temperatures in several layers or as a grid. According to the philosophy of the manufacturer a "homogeneity index" is calculated from the recorded measurements. However we have to assess this grid system critically because of the influence of the burners shown a clear effect on the upper elements and the aging of the thermocouples must be considered.

More than 30 years ago extensive research studies with respect to the glass temperatures inside the spout took place [7]. Nowadays it is not possible to perform such studies in the normal production because it is expensive and much care by man-power is needed. The previous research results were increasingly replaced by simulations [8, 9]. As a result, the feeder bowl geometry could be successfully improved. Nevertheless, the real processes within the spout are still not clearly analyzed.

An actual project is intended to link the temperature distribution of the 9-point grid measurement and the temperature distribution of the gob after the bushing ring by means of mathematical modeling. The data should be linked in a meaningful way and should allow conclusions regarding changes. At press time, the model based interpolation includes the glass, the tube, the atmosphere above the glass, and the refractory material of the feeder. The tube rotates, the movement of the plunger and the shear cut are not considered so far. The outlet is formed as a double-gob.

The first step was to model a well-known system of a feeder channel. This was done with the data of Greschat [7].

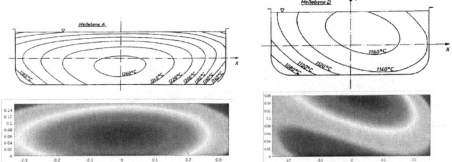

Figure 8: reconstructed isotherms at different cuts of a feeder [7] and first results of FEM-calculations based on the experiments of Greschat

In the next steps the initial- and boundary conditions were improved. So far, measurements were taken at 2 feeders (flint glass and amber glass). The first results show qualitative correct progressions. The well-known difficulties are due to the correct fitment or modulations of the parameters for the calculations. This involves the data of the used media, most taken from literature. In detail these are the thermal conductivity, heat capacity, expansion coefficient and the density of the glass, the air and all the refractory elements. The speed of the glass stream is back-calculated based on the mass flow. Yet, integration of the process specific parameter for the heating has to be improved.

At press time results for an amber glass feeder are already available. Basic data of layer 1 (temperature distribution at the grid measurement) are given in Figure 3. Starting from a nearly symmetrical temperature distribution at the grid measurement with a direction of rotation of the tube clockwise the following results can be presented. A considerable shift of hot regions due to the tube rotation is shown in Figure 9.

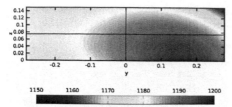

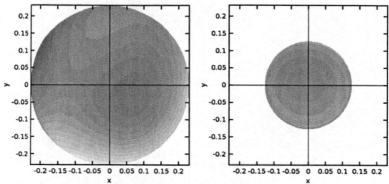

Figure 9: temperature distribution before entry to the spout, layer 12

Immediately below the tube the inflow of hot glass depending on the rotational direction of the tube is visible. Since the tube acts as an heat sink there must be a local cooling at the position of the inflow. This is lost on the way to the orifice ring. If the plunger is also aligned centric, there should be a symmetrical temperature profile at the over top of the orifice. Both are seen in Figure 10.

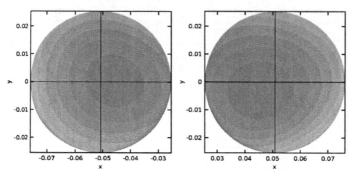

Figure 10: temperature profiles inside the spout at layer 13 and 14

This leads to temperature gradients in the plane of the orifice shown in Figure 11:

Figure 11: result of the calculated temperature distribution of the gobs as horizontal cuts (pos 17, 18)

The hot regions are vis-a-vis, which is in agreement with the measuring results in Figure 5.

The calculated heat transfer coefficients in the delivery are between 900 und 1500 W/(m² K). As a result of an experiment with a changed amount of scoop-spraying a change of the heat transfer coefficient of about 10 % was observed.

The Figure 12 shows both measured and calculated data from one campaign with different depths of visibility by using different wavelengths (pyrometers). In this picture only the part of the gob is shown which was in contact with the trough. This good agreement between measured and calculated data is obtained by applying a heat transfer coefficient of about 1200 W/(m² K).

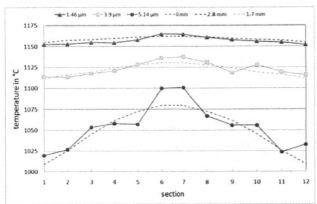

Figure 12: measured and calculated data of the gob temperature at the end of deflector

The elongation of the gob on his way between the shear cut and the blank was measured. The FEM calculations were carried out so far, however, section by section and with a rigid model. This results in discontinuities in the interpolation, because of a change of length is always connected with a reduction of the diameter. After elongation the recorded temperature now is a signal out of another part of the gob (with the same depth of visibility). One of the first steps of a planned follow-up project will be the change of the model from rigid to dynamic, in order to ensure that the colder regions are less deformed than the hotter regions.

The diagram in Figure 13 summarizes the results of interpolations using the example of one machine section. A reduction of the temperature gradient takes place between shear cut and the scoop. The cause is given by the release of energy by radiation from the near surface area of the gob firstly, and a tracking of energy out of the gob center secondly. The temperature gradient of the gob becomes steeper in contact with metal parts. This means that periods of reheating are quite essential for a good temperature distribution at the fall into the blank.

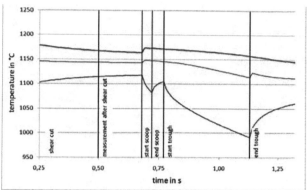

Figure 13: interpolation of the development of temperature distribution of brown glass at one section in dependence of time

SUMMARY

The temperature distribution of the gobs falling into the blank is a result of the temperature distribution in the feeder channel and the change during passage through the spout on one hand. On the other hand the distribution is also influenced by the interaction with the delivery and the environment. Therefore an optimization of an IS-machine requires the understanding of all sub-processes, and the ability to realize specific temperature profiles between feeder and blank is needed.

At the delivery the model-based interpolations show a very good agreement with the determined data for the previous measurements. Based on the results achieved so far, the interpolation is now extended to the more complex area of the spout.

ACKNOWLEDGMENT

The here shown results are based on research financed within the agenda for promotion of industrial collective research (IGF) by the Bundesministerium für Wirtschaft und Technologie via the Arbeitsgemeinschaft industrieller Forschungsvereinigungen (AiF). The authors wish to thank the management and staff of the Gerresheimer Lohr GmbH and Verallia AG Bad Wurzach.

REFERENCES
[1] Zimmermann, H., Merkwitz, M., Strack A.: Die Bedeutung der Viskosität für die Konditionierung und Verarbeitung von Glasschmelzen, Glas Ingenieur, 2000, No.1, p. 39-43
[2] Manns, P.: Untersuchungen zum Klebe- und Abriebverhalten von Formenwerkstoffen für die Glasheißverarbeitung, AiF-Forschungsvorhaben Nr. 13508N, 2005
[3] Lochner, K.; Raether, F., Bergmann, G.: Verfahren zur präzisen Viskositätsregelung am Glastropfen. AiF-Forschungsvorhaben Nr. 14390N, 2007
[4] Größler, J.:Thermal Homogeneity Index – the real truth, Nikolaus Sorg GmbH & Co KG, 2007
[5] Krause, D., Loch, H.: Mathematical Simulation in Glass Technology, Springer-Verlag, 2002
[6] Nölle, G.: Technik der Glasherstellung, Deutscher Verlag für Grundstoffindustrie, 3. Auflage, 1997
[7] Greschat, K.-H., Meister, R.: HVG-Kurs Optimierung von Glasschmelzöfen, 1981
[8] Franzel, H.: Strömungs- und Temperaturverhältnisse im Speiserkanal von Hohlglasmaschinen, Teil 1, Glastechn. Ber., 1975; No.2; p. 27-34
[9] Hyre, M.R., Paul, K.: Feeder design and gob shape, Glass Technology, 2000, No.9, p. 18-28

HIGH VISCOSITY GLASS SHEET FABRICATION

Daniel Hawtof
Corning, Inc.

ABSTRACT

A novel apparatus and process for making thin, continuous, high viscosity glass soot sheet and sintered glass sheet is described. Glass soot particles are deposited onto a curved deposition surface of a rotating drum to form a soot sheet. The soot sheet is then released from the deposition surface. The soot sheet is conveyed to a heating source and sintered into a consolidated glass ribbon. The nominal thickness of the glass sheet is 100 microns. The soot sheet and the sintered glass can be sufficiently long and flexible to be reeled into a roll. Materials enabled by this process include silica sheet and doped silica sheet, in single or multiple layer configurations.

BACKGROUND

Glass sheets have been fabricated by many means for generations (1). Float processes were made viable in the 1960's and are used in the formation of glass on a molten metal bath. Glasses obtained by this method are varied in composition range and are used in a wide array of products, such as in windshields, constructions, and other plate glass areas. Slot draw was invented in the 1940's and provides glass pulled with gravity and edge rollers from a melt through an orifice, touching the glass surface with the exit orifice. Fusion glass processing was invented in the 1960's and used to make continuous ribbons or sheets of glass without touching the external glass surfaces. While first anticipated to compete with float glass in conventional plate glass applications, this process and the glass quality enabled by it came to the forefront when the TV and computer screen applications made use of the inherent smooth surfaces and glass composition attributes, such as thermal expansion coefficient. The glass produced in fusion processes is of a melting temperature low enough to enable the molten glass to flow over an "isopipe" and "fuse" together as it is formed. This temperature is in the sub-1000°C range. Over time, products and processes have been evolving from small sheets of relatively thick glass to larger sheets of ever thinning glass. One mm thick glass sheets have yielded to 0.7 mm to 0.5 mm and demonstrations of thinner yet have been performed at 0.2 mm and even 0.1 mm (2). Paper sized sheets have been eclipsed by generation after generation of scale-up in fusion, to multi-meter sizes of sheet used in LCD fabrication. However, fusion processing is limited to relatively low melting temperature glasses that can be flowed over a ceramic isopipe. A new method for making high melting point glass in a thin, roll-able ribbon that can withstand high post-processing temperatures has been invented (3). The glasses that are made possible in this process are neither fusion formable nor practical in float processing. Pure silica, doped silica, and multiple layers of these glasses are made using this new method. The format is in a nominal 100 micron thickness of glass. The current 25-100 mm width is scalable with equipment modifications to wide sheet. Both discreet sheets and continuous ribbons have been demonstrated.

PROCESS OVERVIEW

The silica glass (or doped silica) is fabricated using flame hydrolysis techniques, invented by Dr. Frank Hyde (4) that has been used for fabricating glass for optical fiber. A burner uses methane and oxygen that are premixed to produce a flame. This flame is supplemented with additional burner ports that provide additional oxygen to fully combust the glass precursor that is also delivered through the burner. A typical glass precursor used in this process, as in optical fiber manufacturing, is an organometalic siloxane – octamethylcyclotetrasiloxane (OMCTS) (5). This material is ideal as a silica precursor as it contains silicon, oxygen and methyl groups – so it wants to burn on its own with an ignition source. The OMCTS is vaporized in a glass bead vaporizer at 190°C and delivered with a carrier gas of nitrogen to the burner. The flame and glass precursor are combusted to form silica particles, called "soot". These particles are about 100 nm in size and can be collected on a target substrate.

The target substrate is a quartz drum that is rotated by the burner and is close to the width of the burner. The soot particles collect on this drum and themselves as the surface of the drum moves past the flame. In this way, a sheet of soot is created on the drum surface. The flame flows, distance of the burner to the drum, and drum speed impact the density and thickness of the soot sheet that is created. After the soot sheet is deposited to the desired thickness and density and moves past the deposition zone, the soot sheet is released from the drum surface. At this point, the soot sheet becomes a self-supported body and can be gripped by its edges while one end is still attached and being continually created on the drum.

The soot sheet can be directed by its edges to a series of rollers/tensioners that grip the edges of the soot sheet and tension the sheet to be flat. While the soot sheet is supported and tensioned to be flat by the edge rollers, a heat source is directed to the center area of the soot sheet. Obviously, the heating of the center portion of the soot sheet has to be well enough contained to not overheat the edge rollers that are tensioning the sheet. This is challenging to the extent that the heat required to sinter the soot sheet into fully densified and clear glass is on the order of 1500°C. A variety of heat sources can serve this purpose. Induction heating has advantages for uniformity and low velocity. A hydrogen and oxygen torch has advantages in heat transfer and ease of starting and stopping the heating process. Lasers, resistance heaters, and other methods have similar capabilities and particularities. The goal is to increase the soot sheet density to fully sintered glass (or an intermediate density target) as it is maintained in a flat orientation. In this way, the only shape change during volume reduction of the soot sheet is in thickness. A typical reduction in soot sheet thickness is from 400 microns soot sheet to 100 microns glass ribbon.

Once a portion of the width of soot sheet is converted to sintered glass, the now composite of soot and sintered sheet is directed to CO2 lasers for trimming. Carbon dioxide lasers ablate silica glass and trim the soot sheet edges from the sintered ribbon of glass. Free of the soot edges, the thin ribbon of glass is either cut to a desired length or wrapped onto a take-up drum.

SOOT SHEET DEPOSITION

The soot sheet is made up of particles of glass that have a relatively high volume to mass – the particles look like snowflakes under high magnification. These soot particles have a nominal diameter of 100 nm, are roughly spherical, and have a rough surface structure. Soot particles will agglomerate together to form larger bodies in the time that they travel in the soot stream generated by the burner. The heat in the burner flame partially melts the particles together in the fume stream as well as further heating the particles when they are adhered to the target and to other particles.

Several factors thus determine the soot sheet density. These factors include the flow rates of glass precursor and other burner gasses, the ratio of glass precursor to other burner gasses, the distance between the burner and deposition target, and the duration of time that the deposition target is in the burner fumestream.

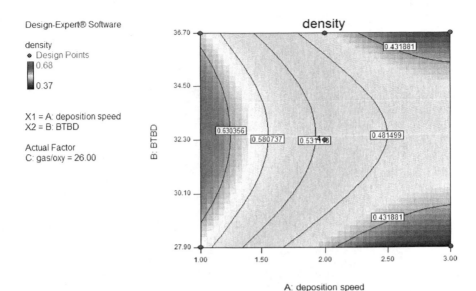

Figure 1 Soot sheet density as a function of distance and speed through flame deposition

Figure 1 shows the density response to varying the speed of the target drum which in turn varies the linear velocity of the resulting soot sheet. The X-axis is the linear soot sheet velocity in mm/second. The term "BTBD" abbreviates "burner to bait distance" and is the distance from the face of the burner to the target drum. In this data, the premix flame was run at a total of 26 standard liters per minute. The density (g/cc) is seen to increase to a maximum and then decrease as BTBD is increased. This is due to the rate of soot deposition changing with distance as well as density. At the highest rate of deposition, the density reaches a maximum as well. Closer to the target, the rate is less since particles have not agglomerated as much. Farther from the target than optimum, the rate decreases as particles cool and there is less driving force from thermophoresis for collection. The variation of density with speed is straight forward as a function of time in the flame. Just as a finger may be moved through a flame at high speed and remain mostly unheated, the same effect is central to controlling the heat absorbed by the soot to modify the soot sheet density.

The burner used to create the soot sheet is a linear design (6), instead of typical optical fiber burners that are rotationally symmetric.

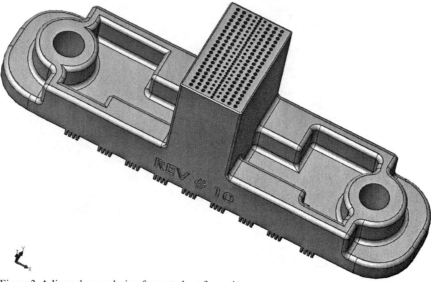

Figure 2 A linear burner design for soot sheet formation

The linear burner shown above has 9 columns of parallel orifices and is symmetric about the centerline. The center column is the only line of orifices that flows a gas that is not split by the center symmetry. The center orifices expel the glass precursor material and optionally a carrying gas, such as nitrogen. The lines of orifice directly on either side of the centerline flow additional oxygen, to provide the proper chemistry for full combustion of the glass precursor material and the premixed flame. The next two sets of orifices do the same. The outside two columns split the delivery of the premix methane/oxygen flame that ignites the glass precursor and provides heat for the reaction and partial density control. The burner was fabricated to enable a series of burners to be mounted together in a linear chain, although a single, longer burner can and has been made as well.

SOOT SHEET SINTERING AND LASER TRIMMING
Several sintering methods have been used and others are available. One way that the soot has been sintered has been to use induction heating. In this case, an induction power supply was used to couple energy into a graphite susceptor and the heat from the susceptor was radiated to the soot sheet to densify the sheet. The schematic shown below in Figure 3 depicts main components of the system. In this figure, the deposition target drum is shown and labeled as 120. This schematic depicts two linear burners depositing soot – one onto the drum (110) and then a second burner depositing on the soot sheet once the soot sheet has been released from the drum. The second burner (116) is depositing on the side of the sheet that was in contact with the drum. The edge guiding and tensioning system (130) serves to move the soot sheet and tension it to be flat as the sintering part of the process is performed. The heating source (140), in this case induction heating susceptors, is shown on both sides of the soot sheet. Heating to fully densified glass has been accomplished from one side as well.

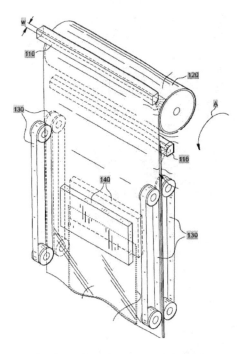

Figure 3 Schematic of soot sheet formation and sintering

The goal of tensioning the soot sheet during sintering is to keep the sintered glass in the same plane as it shrinks in volume, in essence, to control the shrinkage to be primarily if not fully, in the thickness axis. A picture of a sintered section of silica glass, still in the surrounding soot sheet on its edges is shown in Figure 4, below. The thickness of the sample of silica glass shown is nominally 100 microns and the width is about 50 mm. Various thicknesses have been produced by both controlling the starting sheet thickness and alternatively, stretching the sintered glass. The laser trimming, shown below as well, is performed in-line with sintering to produce a ribbon of glass suitable for making samples or to roll onto a mandrel. There are coatings, protective material interleaving, and other post formation processes that can be performed on the glass before or during the wrapping operation (not shown)

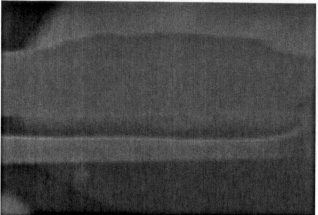

Figure 4   Sintered silica ribbon with soot sheet edges still attached

Figure 5 Laser trimming of sintered ribbon from soot sheet edges

EXPERIMENTAL RESULTS (from ref (3))

A single-layer soot sheet of high purity fused silica was produced using OMCTS as the glass precursor. An example of the linear burner used is showed below. Five linear burners were mounted next to each other, within the distance of the hole spacing pitch on each burner, to create a uniform soot stream the width of the five individual burners. The burners were each 1 inch wide, so a five inch soot stream was produced. The gasses flowed through the burners included: approximately 5 grams/minute OMCTS carried by 20 SLPM of N_2 carrier gas in the centerline orifices of the burners. These gasses were surrounded along their length on both sides by a row of O_2 orifices that provided approximately 5 SLPM of O_2. Outside of these gasses were 2 more rows of orifices providing an additional 20 SLPM of O_2. The final row of orifices, outside of the two, provided a flame to ignite the OMCTS. The flow rates were 12 SLPM of CH_4 and 10 SLPM of O_2.

The burners were positioned approximately 4 inches from the deposition target. The target was a cylindrical quartz drum of 15 inches diameter. The drum had a wall thickness of 0.25 inches. The drum was rotated to provide a surface speed of 1 mm/sec. The soot from the linear array of burners was directed to the drum and a layer of soot approximately 200 microns thick and 6 inches wide was deposited on the drum. The extra inch of soot sheet width was due to flow of the particles along the drum surface). The average density of the soot sheet in the 5 inch wide length of the burners was approximately 1.1 g/cc. The soot outside of this length had a lower density, as it was not in the direct deposition zone of the burners. The 5 inch wide soot sheet created at the higher density was released from the drum, augmented by a stream of air supplied by an air knife. The air knife supplied approximately 20 SLPM of air through a 10 inch wide air knife body, directed at the drum. Besides helping to release the soot, the air knife kept the drum temperature uniform and cool, at about 200° C. The 5 inch wide soot sheet was manually grabbed by the peripheral edges and directed to a wind drum. The drum was approximately 6 inches in diameter. Five meters of soot sheet were wound onto the drum before the experiment was ended.

Another soot sheet with a thickness of 60 microns was fabricated. Three of the above described burners were used in the burner array. Thickness was decreased by increasing the rotation speed of the drum and lowering the OMCTS flow rates. Thickness, density, and production rate of the soot sheet were changed by adjusting burner flow rates, drum rotation speed and distance from the burners to the drum.

A sample length, approximately 2 feet long and 3 inches wide, was taken for a sintering experiment. The peripheral edges of the soot sheet were pinned between rollers in contact along the length of the sample. A heat source was provided to sinter the soot sheet. The temperature of the soot reached approximately 1500° C. and the soot sheet densified to clear, sintered glass. The sintered glass was approximately 30 microns thick.

The sintered sheet with un-sintered peripheral edges was removed from the gripping mechanism and the edges were trimmed off. A 5 watt laser was used to trim the un-sintered soot from the sintered sheet. The laser was traversed at approximately 3 mm/s along the length of the sheet. Both sides were trimmed off in turn, although 2 lasers could have been employed at the same time.

REFERENCES
(1)   11 Mar. 2010, Ellison, A, Cornejo, I, Applied Glass Science, Vol. 1, Issue 1, pp. 87-103.
(2)   8 Aug. 2011, News Release, Nippon Electric Glass Co., Ltd, "NEG Delivers Ultra-thin Lightweight Mirrors for JAXA's Space Solar Power Systems".
(3)   16 Mar. 2010, Hawtof, D., Brady, M., US-7677058 B2, "Process and apparatus for making glass sheet", Corning, Inc.
(4)   Nov/Dec 2000, Keck, D., "A Future Full of Light", Selected Topics in Quantum electronics", IEEE Journal of Quantum Electronics, Vol. 6, Issue 6.
(5)   27 Aug. 1991, Dobbins, M, McLay, R., US-5043002, "Method of Making Fused Silica by Decomposing Siloxanes", Corning, Inc.
(6)   25 Mar. 2008, Hawtof, D., Peterson, R., et al., US-20100143601 A1, "Three Dimensional Micro-Fabricated Burners", Corning, Inc.

CLOSED LOOP CONTROL OF BLANK MOLD TEMPERATURES

Jonathan Simon and Braden McDermott
Emhart Glass Research Center
Windsor, CT, USA

ABSTRACT

In the production of hollow glass containers, the blank mold temperature influences the thermal state of the parison, which in turn influences the final container quality. Traditionally this important process parameter is adjusted manually and the results depend upon the skill and diligence of the operator. To provide more consistent production, a practical automatic control of this key process has been developed. The challenges that were faced, the approach taken and experimental results of successfully applying the new control are presented.

INTRODUCTION

Commercial hollow glass containers are typically produced in an IS (Individual Section) Machine that utilizes a two stage forming process. In the first stage, known as the *blank side* process, a hollow preform (parison) is formed, by either pressing or blowing a gob of molten glass within the cavity formed by a pair of *blank mold* halves. In the second stage, known as the *blow side* process, the parison first reheats from its hot interior and stretches under the influence of gravity. It is then blown using compressed air within the cavity formed by the two halves of the blow molds to form the final container shape. The shape of the outer surface of the finished container is determined by the dimensions of the blow mold cavity. In contrast, since the inner surface is formed by compressed air, there is no predetermined shape for the inner surface. Thus, the wall thickness depends upon the redistribution of glass that occurs in the blow side process as the container transforms from a parison into a finished container. This, in turn, depends upon the highly temperature dependent viscosity distribution of the parison as it enters the blow side process. For this reason, it is essential to obtain parisons whose thermal state is consistent over time, and across the multiple cavities of the overall machine. As the temperature of the blank side mold equipment has a primary influence on the thermal state of the parison, it can be understood that it is important to maintain the desired values for this process parameter.

Today, despite the importance of maintaining proper blank mold temperatures, the cooling for these molds is typically adjusted manually by the operator. Due to disturbances, such as changes in cooling air temperature that affect the system the required amount of cooling varies over time. Operators therefore not only need to setup the initial cooling timing, but a must also readjust it during each shift in order to maintain the desired blank temperatures over time. This manual adjustment is typically done without the benefit of on-line measurements, based upon observation of the glass itself and spot checks of the blank mold temperature using hand held probes. The results are thus highly dependent upon both the skill and diligence of the operator.

A new approach is described here, utilizing a closed loop control where the temperatures of the blank molds are measured and the cooling is automatically adjusted to maintain a desired temperature value.

167

BLANK MOLD THERMAL PROCESS

To better understand the operation of the closed loop system and the approach that was taken it will be beneficial to first provide a more in depth description of the blank mold thermal process. On the blank side, heat is transferred to the interior surface of the blank molds from the hot glass and removed by cooling air which is passed through cooling passages within the blank molds. The process operates cyclically. In each machine cycle, a fresh gob of molten glass loads into the closed blank mold halves, it is then transformed into a parison, the blank mold halves are opened and the parison is removed, and finally the blank mold halves are closed and made ready to accept the next gob.

Following the basic machine cycle, both the heat addition and removal from the molds are also cyclic. In fact, during each cycle, the interior surface switches between two modes. In the first mode, there is hot glass in contact with the mold surface with a large heat flux into the mold. In the second mode, the glass has been removed and a relatively small amount of heat flows out of the interior mold surface due to natural convection and radiation. During the machine cycle, the cooling air supply valve is switched between open and closed providing a pulse of cooling air flow through the internal cooling passages of the mold. The quantity of heat removed per cycle is adjusted by controlling the duration of the pulse. Thus, the heat transfer condition within the cooling passages also switches between two modes. In the first mode, the coolant is flowing and there is a rapid removal of heat from the cooling passage surfaces. In the second mode, the flow is stopped, and there is a negligible amount of heat transfer.

The results of the cyclic heating and cooling are illustrated in Figure 1 and Figure 2, which shows the predicted temperatures within the glass and blank mold, respectively, obtained from our simulation of a one dimensional unsteady model of the blank mold thermal process. Overall, the periodic nature of the response is quite evident, and also the sharp spatial temperature gradients are also apparent, indicating that at these time scales, the mold metal cannot be regarded simply as a single lumped thermal mass with a uniform temperature. Looking in further detail, and first considering the glass (parison), it can be seen that after the gob makes contact with the molds the glass surface temperature drops rapidly and a sharp temperature gradient or thermal skin forms on the parison. After the molds open and contact is broken, this skin then begins to reheat from the hot interior of the glass. Looking at the mold, it can be seen that when the glass comes into contact with the mold, the temperature of the mold surface that is contact with the glass rapidly increases and then the elevated temperature penetrates into the interior surface of the mold. After the glass is removed, the surface temperature drops, as heat continues to be conducted from the surface into the interior. Along the cooling passages, the mold surface temperature decreases after the coolant is switched on and then warms back up as heat from the relatively warm interior of the mold conducts back to the surface.

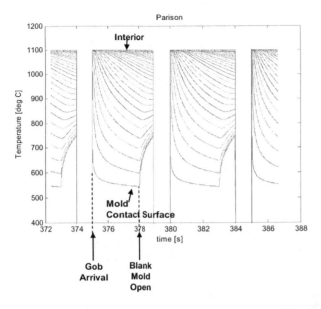

Figure 1 Parison Temperature Variation with Time and Radial Location

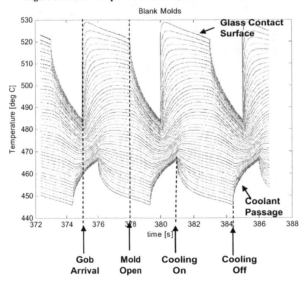

Figure 2 Blank Mold Temperature Variation with Time and Radial Location

CHALLENGES OF AUTOMATICALLY CONTROLLING BLANK MOLD TEMPERATURE

A number of challenges had to be surmounted in order to implement a practical and effective automatic (closed loop) blank mold temperature control system.

A prerequisite for any closed loop (feedback) control system is a means to measure the parameter that is to be controlled, in this case, blank mold temperature. In general, it is difficult to implement sensors and measurement systems that will survive in the harsh, high temperature conditions found in the glass container making process. In particular, a typical IS machine has 10 to 12 sections each with between 2 and 4 blank mold cavities per section, each cavity consists of two mold halves thus there can be between 40 and 96 blank mold temperatures to be measured in each machine. Furthermore, the mold halves move as the molds open and close and the molds are replaced on a regular basis. Implementing typical sensors such as thermocouples in this situation would be quite difficult, due to the routing of the large number of cables that would be required, providing interconnects for changing molds, and allowing for flexing of leads as the molds open and close. In this situation, a non-contact measurement is certainly preferred; however, the expense of providing individual, non-contact measurements for so many measurement points would appear impractical. Implementation of a practical closed loop control system has been enabled by the availability of a commercial system, the Emhart TCS ™ System shown in Figure 3, which has been developed to provide blank mold temperature measurements using a single pyrometer, and a 3 axis mount. In this system the pyrometer traverses the entire length of the machine and is then aimed at individual stations to obtain measurement of the temperatures of the individual mold cavities. The pyrometer continuously traverses back and forth providing a periodic update of each temperature measurement point.

Consideration of the underlying process to be controlled reveals that the closed loop control of blank mold temperatures differs in a number of key aspects from the typical industrial process control problem. In the typical process control problem, the dynamics of the process to be controlled, e.g. the level in a tank, or the temperature of a well mixed flow, are well represented by a low order, lumped parameter (ordinary differential equation) dynamic system with continuous (in time) inputs from the actuator used to adjust the process. In contrast, as previously discussed, at these time scales, blank molds clearly exhibit the characteristic of a distributed parameter system, in which the temperatures vary spatially in the radial (depth into the mold), circumferential, and axial directions. Also, both the control input (cooling air), and heat inputs (hot glass) are discontinuous in time and switch periodically with each successive machine cycle of gob loading and parison formation. These factors make the control design problem somewhat more challenging as it is not possible to directly derive simple analytical process models to be used for analyzing and designing the control. Instead the control development must rely upon more elaborate simulation models, and experimental testing with the actual process.

Finally, in order to implement a closed loop control system it is necessary to have a means to adjust the timing of the start and/or end of cooling air pulse, and maintain these values within defined limits relative to the other system timing events. It is also important to have additional information regarding the state of the machine sections being measured, for example whether they are running, or have recently been swabbed (Operators periodically apply lubricant to the molds in an operation known as swabbing). In general, commercial timing control systems do not provide such an external interface. To overcome this issue, the blank temperature control system described here, is being integrated directly within the Emhart FlexIS™ timing control system to provide the necessary intercommunication between the control and timing system.

Figure 3 Three Axis Traversing Pyrometer

CLOSED LOOP SYSTEM DESCRIPTION

The basic structure of the closed loop temperature control system is shown in schematically in Figure 4. As indicated in this figure, for each cavity, the setpoint (desired) temperature is compared with the measured value from the pyrometer. The difference is operated upon by a control algorithm which then computes an adjusted cooling duration. The adjusted cooling duration is then applied to the physical blank molds by the timing control system, and the new resulting temperature is measured by the pyrometer completing the loop.

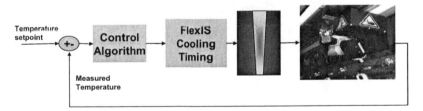

Figure 4 Closed Loop Control Structure

As previously discussed, the temperature of the blank molds varies both spatially and temporally. It is noted that the measurement made by the traversing pyrometer is made at a single location on the blank mold surface and at a single point in time relative to the machine cycle. For example, the temperature can be read just after the glass leaves the blank molds as soon as the mold surface can be seen by the pyrometer. The system is inherently a sampled data system; with the sampling period set by the round trip time required by the pyrometer to measure temperatures for the entire machine. Typically this may be between 5 to 15 minutes depending upon the number of measured cavities and the cycle rate of the machine.

CLOSED LOOP EXPERIMENTAL RESULTS

The closed loop system has now been successfully implemented and tested on operating IS machines. Some key results illustrating the capabilities of the closed loop system are presented in Figures 5 - 7. The command following capability of the closed loop system is illustrated in Figure 5 for four different mold cavities placed under automatic control. At t=0, the temperature setpoint is changed from 465 °C down to 445 °C. It can be seen that the four cavity temperatures move and settle to the new commanded setpoint value, without excessive overshoot or oscillation. This demonstrates how an operator can directly modify the cavity temperature simply by adjusting a setpoint value. In contrast, without an automatic control, even if measurements were available, the user would have had to adjust the cooling time by trial and error until the desired temperature was achieved. Since the open loop system typically takes approximately 20 minutes to completely respond to each change in cooling duration, this trial and error adjustment can be lengthy, resulting in extended periods in which the production quality may be adversely impacted. The difficulty, in manually adjusting the cooling duration is further reinforced by Figure 6, which plots the controller outputs (cooling durations) that were required to obtain the temperature changes shown in Figure 5. It can be seen that the actual final magnitude of the required adjustment is somewhat different for each cavity. It is noted that the required amount of cooling will also depend upon other factors such as the cooling air temperature, and fan pressure, further motivating the use of an automatic control to provide accurate and timely changes to the mold temperatures.

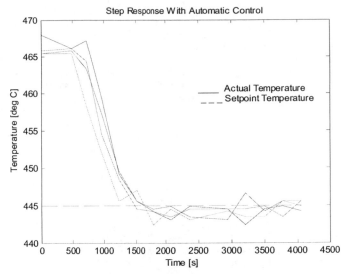

Figure 5 Command Following Step Response

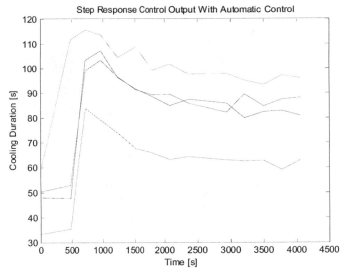

Figure 6 Step Response Control Output

In addition to the ability to obtain quick and accurate command following, another major benefit of the closed loop control is the ability to maintain constant cavity temperatures in the face of unavoidable disturbance inputs. Typical disturbances include changes in cooling air temperature and supply pressure. To assess the capability of the closed loop control to maintain a constant temperature in spite of disturbances, a test was conducted during a field trial on a 12 section triple gob machine. In this test half of the sections were placed under automatic control and the other half were left in manual. Specifically the cooling durations for the odd numbered sections were adjusted by the automatic control and the even sections were left in manual, so their cooling times were not adjusted. The results are shown in Figure 7 which plots the deviation of the blank mold temperatures from their initial values for the two cases. It can be seen that over the 10 hour test period the temperatures of the manual sections keep drifting further from their initial values. By the end of the test they range by +/- 15 °C. In contrast, the sections under automatic control maintain their temperatures within a relatively tight band typically +/- 5 °C. In Figure 8, the cavity temperatures and controller outputs are shown for one of the automatically controlled sections for this same test. It can be seen that in order to maintain the constant temperatures, the cooling duration must be continually adjusted. It can be appreciated that without the automatic control it would have required a great deal of time and attention on the part of the operator to make such adjustments, and further, that it would be difficult for the operator to know the precise amount of correction that would be required. The cyclic disturbances evident in both the automatic and manual data shown in Figures 7 and 8 is due the periodic application of lubricant (swabbing) of the molds that is performed approximately every 30 minutes. It can be seen that the automatic control can also handle such disturbances, and returns the temperatures to their setpoint values.

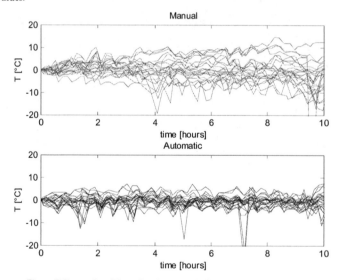

Figure 7 Comparison Manual and Automatic Control

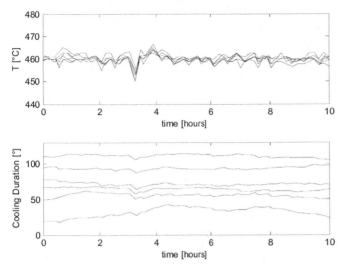

Figure 8 Cooling Adjustment to Maintain Constant Temperature

SUMMARY AND CONCLUSIONS

The blank side thermal process strongly influences the thermal state of the parison, which in turn influences the final container quality. Due to process disturbances and variability in flow distribution, the amount of blank mold cooling required in order to maintain a consistent thermal state of the parison varies over time and for different mold cavities. Traditionally, to compensate for this variability, the amount of cooling is adjusted manually and the results depend upon the skill and diligence of the operator. To provide more consistent production, a practical automatic blank mold temperature control system has been developed.

The system that has been developed has been enabled by the combination of: 1) a traversing pyrometer providing the necessary measurements, 2) a close integration with the timing control system allowing the machine cooling durations to be directly adjusted, and 3) an understanding of the system dynamics gained from analysis and simulation of the underlying physical process.

The system has been experimentally evaluated on operating IS glass production machines, and the results show that the automatic system is capable of providing automatic adjustment to the operating temperature, and maintaining a consistent production temperature.

Adoption of this new technology can be expected to reduce the variability of the blank mold temperatures ultimately improving overall yields. By relieving operating personnel of the need to constantly adjust the blank cooling, it can also be expected that they will have more time to beneficially devote to problem solving in other areas.

Legislation, Safety, and Emissions

THE U.S. POLICY AND POLITICAL LANDSCAPE AND ITS POTENTIAL IMPACTS ON THE GLASS INDUSTRY: THROUGH A GLASS DARKLY

Shelley N. Fidler and Marisa Hecht
for Van Ness Feldman & Air Products and Chemicals

INTRODUCTION

Against the backdrop of a turbulent economic climate and entrenched partisan divisions in Congress, the policy and political landscape affecting the energy sector and energy-intensive industries has seen dramatic changes in recent years. With the glass manufacturing's energy-intensity, legislative and regulatory activity in Washington necessitates close attention from the industry.

The 2010 Congressional elections dramatically changed the energy debate in Washington. Congress turned its focus away from comprehensive climate change and clean energy legislation to battles over fiscal policy and deficit reduction. To the extent Congress addressed energy and environmental issues, its attention was limited to attacks on regulation, reduction in incentives for clean energy development, and attempts to increase fossil energy production. Meanwhile, the Obama Administration has relied on its existing legal authorities to advance its energy and environmental policy objectives. On the regulatory front, the Environmental Protection Agency (EPA) issued rulemakings, primarily according to a schedule required by law or by settlement agreements to legal challenges that changed the economics of coal and gas electricity generation. The Department of Energy (DOE) worked to advance clean energy research and development and encouraged transmission projects. And the Federal Energy Regulatory Commission (FERC) instituted reforms to promote transmission planning and investment, the smart grid, energy storage, and demand response.

CONGRESSIONAL ACTIONS

Congress issued the Budget Control Act of 2011 in August 2011 which resolved the debt ceiling impasse. The Act provided nearly $900 billion in Federal spending cuts over 10 years. It also established a new Congressional committee, the Joint Select Committee (JSC) on Deficit Reduction (known in Washington as the "Supercommittee"), charged with developing recommendations for at least $1.2 trillion in additional deficit reduction – an ambitious project that collapsed in November 2011 as both parties failed to reach agreement on spending cuts or tax increases. As a result, starting in 2013 the federal government will have to implement a total of $1.2 trillion in automatic spending cuts over the next ten years. A $2.1 trillion increase in the debt ceiling is expected to carry through calendar year 2012.

There are also other deficit reduction proposals that will potentially impact the fiscal climate in Washington. They remain options to be chosen by policy makers if they desire other deficit or spending reductions not already agreed upon. These proposals include the President's Commission on the Fiscal Responsibility and Reform (known as the Bowles/Simpson Commission); the Bipartisan Policy Center's Debt Reduction Task Force (Domenici/Rivlin); the House Republican FY2012 Budget Resolution; the Senate Gang of Six Deficit Reduction Plan; and the Congressional Budget Office's Reducing the Deficit report. At present, none of these proposals appear likely to be enacted into law. However, with the automatic spending reductions looming and the Bush-era income tax cuts scheduled to expire at the end of 2012, Congress may be motivated to attempt further compromises on fiscal policy over the coming year. Funding for energy programs, and the continuation of energy-related tax credits and incentives, is widely expected to be an important element of those discussions.

REGULATORY ACTIVITIES

EPA has issued several rules, and has a number of rulemakings under development, that may affect the glass manufacturing industry due to its energy-intensive nature. These rules include direct regulation of industrial emissions through the Boiler MACT/GACT rule and potential revisions to National Ambient Air Quality Standards (NAAQS). Furthermore, EPA rules that regulate the power sector, such as the Cross-State Air Pollution Rule (CSAPR) and the Utility MATS, may have indirect effects on the glass manufacturing industry. All of EPA's rules, whether having a direct or indirect effect on the glass manufacturing industry, may require costly retrofits with add-on controls, may constrain production through tighter regulatory requirements, and may incent efficiency because of higher energy costs.

Section 112 of the Clean Air Act (CAA) requires EPA to set Maximum Achievable Control Technology (MACT) standards for listed source categories and Generally Available Control Technology (GACT) for smaller sources. In February 2011, EPA issued final rules that require reductions of hazardous air pollutants (HAPs) from affected boilers and process heaters at manufacturing sources. MACT standards apply to HAP emissions from affected boilers and process heaters (indirect-fired only) burning various fuels including coal, oil, biodiesel, natural gas, landfill gas, and biomass at major sources.[1] The rule imposes work practices on natural gas boilers and process heaters and emissions limits on other boilers and process heaters. The rule requires all sources with existing boilers and process heaters to conduct a one-time energy assessment or audit to identify cost-effective energy-saving measures that will also decrease emissions – particularly relevant in the high-energy-use glass manufacturing industry. Boilers and process heaters that are already subject to other MACT standards are not subject to the boiler rules. EPA also set GACT standards for those boilers and process heaters at area sources.[2] Natural gas- and landfill gas-fired sources are not covered under the GACT standards. The GACT standards impose emissions limits and work practices on affected sources.

EPA is in the process of reconsidering aspects of these rules, with a final reconsideration anticipated this Spring. In the meantime, EPA attempted to suspend or "stay" these rules while it completed its reconsideration, but the stay was overturned by the D.C. Circuit Court of Appeals in early 2012. Industry and environmental group challenges to the boiler rules issued last year also remain pending. On the Congressional front, H.R. 2250 aims to halt the Boiler MACT and would give EPA 15 months to set a new standard, with industry having 5 years to comply. The House of Representatives voted 274-142 on October 13, 2011 to pass the legislation. However, the Senate is not likely to pass the measure and the White House most likely would veto it.

Sometime during Spring 2012, the revised rules are likely to be finalized. These would likely change emissions limits. Also, there would be a potential for improved achievability and a potential for change in the compliance deadline. Currently, the compliance deadline is 2014 but is subject to change depending on the results of the reconsideration. At present, sources are not required to implement measures. For the Energy Audit, all existing sources will be required to perform a one-time energy audit or assessment to identify energy-saving measures. This is not addressed in the

---

[1] Major sources are those that emit more than 10 tons per year (tpy) of any single HAP or 25 tpy of a group of HAPs in the aggregate.
[2] Area sources are those that emit less than 10 tpy of any single HAP or 25 tpy of a group of HAPs in the aggregate.

reconsideration so is not likely to change. Moreover, litigation may go forward depending on the results of EPA's reconsideration. The D.C. Circuit may order EPA to undergo further review.

Additionally, EPA regulations required by other sections of the CAA have impacts on glass manufacturing. Pursuant to Section 109 of the CAA, for example, EPA must set National Ambient Air Quality Standards (NAAQS) for particulate matter (PM) – and states must submit implementation plans demonstrating how they plan to achieve that air quality standard through regulation of industrial facilities and other stationary sources. Recently, EPA announced it will release a new proposed NAAQS regulating PM this summer. Similarly, EPA has been in the process of revising its NAAQS for ozone for some time; the Administration is now in the midst of implementing an ozone NAAQS that was issued in 2008, and is working to develop a new and potentially more stringent ozone NAAQS by 2013. The PM and ozone NAAQS could lead to new state regulations directed at controlling emissions of nitrogen oxides (NO_x) from industrial sources. In addition, these NAAQS could eventually prompt EPA to impose direct regulation on industrial NO_x emissions that have interstate impacts (a step the Agency has already taken with respect to the power sector through its Clean Air Interstate Rule and the CSAPR).

Some rules specifically regulate the glass manufacturing industry. The Glass Manufacturing GACT includes emissions limits for glass manufacturing furnaces. The Glass Manufacturing New Source Performance Standards (NSPS) rule regulates criteria pollutants emitted by glass manufacturing sources constructed after June 15, 1979. Bear in mind that EPA reviews these standards every 8 years pursuant to its requirements under Section 111 of the CAA, with the most recent revisions occurring in 2000.

Indirectly, certain EPA regulations may have impacts on the glass manufacturing industry. Due to the glass manufacturing industry's high energy demands, any rules affecting electric generating units (EGUs) will affect the industry, particularly if these rules increase the demand for and the price of natural gas. For example, EPA's recently-issued CSAPR rule regulates emissions of NO_x and SO_2 from EGUs in the eastern United States that affect states downwind. Although the implementation of CSAPR was temporarily suspended at the beginning of 2012 as a result of legal challenges, it is likely that either CSAPR or a similar rule replacing it will take effect over the next two to five years. EPA also introduced the Utility Mercury and Air Toxics Standards (MATS), which regulate hazardous air pollutants from EGUs. Section 112 of the CAA requires EPA to regulate HAPs through standards based on the top-performing 12% of facilities. Although this rule does not apply to natural gas-fired utilities,[3] it could cause closures of older and less-efficient coal-fired plants. The combination of the Utility MATS and CSAPR will likely create an incentive to use more natural gas in the power sector, driving up its price. EPA issued the Utility MATS on December 16, 2011 with compliance required by either February or March 2015. EPA estimates the cost of the rules to be $10 9 billion in 2016. Most facilities will have to install Hg, HCl, and PM controls to comply with the final standards. Moreover, litigation may ensue due to a tight compliance deadline. Other regulations include the Cooling Water Intake Rule, the Coal Combustion Residuals Rule, and the Greenhouse Gas New Source Performance Standards.

CSAPR was issued in July 2011. The CAA requires states to develop plans to prevent interference with air quality (e.g., ozone, PM) in other states, and CSAPR is EPA's rule pursuant to this requirement. CSAPR has two main components: emissions "budgets" for SO_2 and NO_x for 27

---

[3] The MACT standard does not apply to natural gas-fired facilities, but the NSPS for EGUs (issued simultaneously as the MACT standard as part of the MATS package) regulates natural gas-fired new sources emitting criteria pollutants.

eastern states and an emissions trading program to ensure each state meets its budget (Federal Implementation Plans). Recently, EPA proposed several new state budgets – i.e. Texas's new budget increased by 29%. Also, EPA relaxed restrictions on interstate trading for two years. The emissions budgets limit annual NO_x, annual SO_2, and have an ozone season NO_x in each state. The emission budgets were intended to take effect in 2012, with some states facing lower SO_2 budgets in 2014; however, the recent stay of CSAPR will likely require EPA to postpone those compliance deadlines. The emissions trading system applies to fossil fuel-fired power plants. There are unlimited intrastate trading allowances but limited interstate trading allowances. CSAPR will most impact coal-fired plants by driving installations of scrubbers, SCR, and low- NO_x burners, pushing the switch to low-sulfur coal and natural gas, and causing reduced dispatch or retirement of coal-fired units. Its expected cost is $2.8 billion per year. As noted above, a number of legal challenges have been filed against the CSAPR and remain to be resolved.

Regarding greenhouse gases, EPA is in the process of developing proposed NSPS that apply to new and existing EGUs constructed after the date announced in the proposed rule. NSPS reflect the "best system of emission reductions." Whereas the federal government implements NSPS for new and modified sources, states implement NSPS for existing sources subject to EPA oversight. A proposal for new and modified EGUs is expected in February 2012, but a timeline for proposed existing sources' guidelines from EPA remains unclear.

Other environmental Acts, such as the Clean Water Act (CWA) and the Resource Conservation and Recovery Act (RCRA) also may have effects for the glass manufacturing industry. A rule issued pursuant to Section 316(b) of the Clean Water Act (CWA) may indirectly impact the glass manufacturing industry by virtue of its effects on the power sector. The CWA requires the "best technology available" to minimize impacts on aquatic life from cooling water intake. EPA proposed a rule for existing power plants and industrial facilities in March 2011, and the final rule is under review at the White House Office of Management and Budget (OMB) with an expected release sometime this summer. The proposed rule affects steam generating units and manufacturing facilities that withdraw from "waters of the United States," including approximately 45% of the nation's generating capacity. The proposed rule would impose impingement and entrainment as controls. Impingement, which would be required 8 years after the rule takes effect, reduces intake velocity and provides advanced screens for fish and shellfish. Entrainment, which would have a flexible compliance schedule, includes a state study process and may require some facilities to install closed-cycle cooling. The rule would cost an estimated $497 million annually. In addition, EPA is required to propose rules for power plant effluent this year. Under RCRA, EPA is required to regulate the disposal of coal ash. The final rule is expected sometime in 2012 and may treat it as either hazardous or non-hazardous waste with impoundment retrofits being the most costly.

The Federal Energy Regulatory Commission (FERC) also has regulations in place that could affect the glass manufacturing industry. Order Number 1000 reforms transmission expansion planning and cost allocation, although much is left to compliance filings and regional implementation. It considers reforms to support the use of advanced technologies, such as fast-response energy storage. It also supports participation of demand response providers in organized markets – a topic FERC addressed in early 2011 in an order establishing rules for the compensation of demand resources in organized wholesale energy markets. To the extent that the glass manufacturing industry participates in demand response programs, FERC's rulemakings on transmission planning and demand response compensation bear continued monitoring. FERC is also conferring with EPA on the reliability implications of EPA's rulemakings.

CONCLUSION

In light of the changing governmental climate, affected entities should be in touch with their environmental compliance team regularly to stay current with the new regulatory requirements. Entities should incorporate sustainability and efficiency in all energy consuming process decisions and new offerings. Furthermore, entities should think about how to document and market their energy efficiency and emission reductions to customers and regulators. Finally, members of the industry may consider ways to inform and influence upcoming regulations by highlighting the economic impacts on the glass manufacturing business. With additional policies and regulations directly and indirectly affecting the glass manufacturing industry, it may be beneficial for industry members to keep informed of the changes in order to best address any impacts.

Key EPA Regulatory Activity for Stationary Sources

VanNess Feldman

	EPA Action	Subject	Date of Proposal (Publication in the *Federal Register*)	Date of Final Agency Action (Publication in the *Federal Register*)	Applicable to New ("N") or Existing ("E") Facilities?
Criteria Pollutants (e.g., SO2, Ozone, PM)	Transport Rule (CAIR Replacement)	EPA regulation to reduce SO2 and NOx levels in the Eastern U.S. to replace Clean Air Interstate Rule (CAIR) that was remanded by D.C. Circuit in 2008.	Revisions October 6, 2011	Expected early 2012	N, E
	New PM2.5 Ambient Air Quality Standard	EPA to evaluate whether to tighten existing (2006) fine particulate standard pursuant to court remand.	Expected 2012	Expected 2012(?)	N, E
	New SO2 Ambient Air Quality Standard	EPA finalized rule to replace current annual and 24-hour sulfur dioxide standards with a more stringent 1-hour standard.	December 8, 2009	June 22, 2010	N, E
	New NO2 Ambient Air Quality Standard	EPA finalized a new 1-hour NO2 standard at the level of 100 parts per billion.	July 15, 2009	February 9, 2010	N, E
	New Ozone Ambient Air Quality Standard	EPA proposed rule to enact more stringent NAAQS for 8-hour "primary" ozone standard, and establish a distinct cumulative, seasonal "secondary" standard to protect "sensitive vegetation and ecosystems."	January 6, 2010	Delayed until 2013	N, E
Hazardous Air Pollutants (e.g., mercury, acid gases)	MACT Rulemakings for Mercury and other HAPs from Utility Boilers	EPA to set Maximum Available Control Technology (MACT) standard for all coal-fired power plant mercury emissions, and a range of other hazardous air pollutants emitted by coal and oil-fired power plants.	Mar. 16, 2011	Expected Feb. 2012	N, E
	MACT Rulemakings for Mercury and other HAPs from Industrial Boilers	EPA to set Maximum Available Control Technology (MACT) standard for industrial boilers.	June 4, 2010	March 21, 2011. Stayed indefinitely as of May 18, 2011. EPA issued proposal in Dec. 2011.	N, E
Waste and Water	Coal Combustion Waste	RCRA rules on disposal of coal combustion wastes – possible treatment as hazardous waste.	June 21, 2010	Expected 2012 (?)	N, E
	Wastewater Discharge Regulations under Clean Water Act (CWA)	Regulation of wastewater discharges from thermal generating units.	Expected July 2012	Expected Jan. 2014	N, E
	CWA Section 316(b)	EPA rule to replace remanded rule for regulating cooling water intake structures at existing facilities.	March 28, 2011	Expected July 2012	E
Greenhouse Gases	GHG Reporting Rules for New Sectors	EPA additions to GHG Reporting Rule (finalized October 30, 2009) for fugitive and vented emissions from oil and natural gas systems, CO2 injection and geologic sequestration, and producers and emitters of some fluorinated GHGs.	April 12, 2010	Nov. 30, 2010 and Dec. 1, 2010	N, E
	"Johnson Memorandum"	EPA revised guidance to defer effective date of BACT and Title V permit requirements for GHG emissions from stationary sources until at least January 2, 2011	October 7, 2009	April 2, 2010	N/A
	"Tailoring Rule"	EPA raised GHG threshold for BACT and Title V for GHGs; phase-in starting in 2011 with sources already subject to PSD and Title V	October 27, 2009	June 3, 2010	N
	BACT and Title V Implementation				
	- *BACT Guidelines*	Guidance on what constitutes BACT for GHG emissions from new and modified power plants.	Nov. 10, 2010	March 2011	N
	- *Biomass Emissions*	EPA issued final rule exempting biomass-derived GHG emissions from PSD and Title V requirements for a three-year period to collect data on lifecycle GHG emissions.	March 21, 2011	July 20, 2011	N
	- *SIP Modifications*	Modification of State Implementation Plans (SIPs) to reflect higher Tailoring Rule thresholds. States continue to modify SIPs and state regulations to reflect Tailoring Rule thresholds.	Sept. 2, 2010	Dec. 3, 2010 (SIP call); Dec. 23, 2010 (FIPs)	N/A
	New Source Performance Standards	GHG emission standards for new and modified power plants	Expected Feb. 2012	Expected 2012	N
	Performance Standards for Existing Plants	GHG emission standards for existing (unmodified) power plants	Unknown	Expected 2012	E
	NEPA Guidance from Council on Environmental Quality (CEQ)	Draft guidance from CEQ addresses analysis of direct and indirect GHG emissions that may result from proposed federal actions, and potential impact of proposed federal actions on climate change.	February 18, 2010	Expected ?	N

Studies of EPA-induced Retirements

Study	Regulations Included	Projected Retirements	Details
EIA AEO2011 April 2011	TR, Mercury	14-18 GW total	The high ends represent retrofit cost recovery in 5 yrs vs 20. "TR, Air Toxics, NO_x" assumes wet FGD & SCR on each unit. Nat gas price below AEO2011 (~$4/mmBtu) brings second case retirements up to 40-73 GW.
	TR, Air Toxics, NO_x	19-45 GW total	
EPA March 2011	TR, Toxics	23 GW total (including 10 GW incremental)	Modeling for Utility Air Toxics Rule (Toxics) proposal; Transport Rule (TR) included in the baseline and not in the incremental retirements.
BPC March 2011	TR, Toxics, Coal Ash, 316(b), NO_x	29-35 GW total (15-18 GW incremental)	Assumes ACI, Fabric Filter and either wet FGD or DSI for Utility Air Toxics Rule. DSI only for units <300 MW with low sulfur coal. Cooling towers if >500 MGD design intake. Stricter NO_x by 2018. Low end of the range results from higher AEO2010 natural gas price.
EEI January 2011	TR, Toxics, Coal Ash, 316(b), NO_x	46-56 GW total (24-34 GW incremental)	Low end estimates reflect availability of lower cost compliance strategies for some units. EEI scenarios that include CO_2 price are excluded.
CRA December 2010	TR, Toxics	39 GW total (includes 6 GW planned retirements)	Assumes ACI, fabric filter, and FGD for Utility Air Toxics Rule. Assumes AEO2010 natural gas price.
Brattle Group December 2010	TR, Toxics	40-55 GW total (34-49 GW 2020 incremental)	Doesn't identify specific assumptions for each rule, but assumes SCR and scrubber on every coal unit by 2015. Cooling towers on all coal units by 2015 for 316(b).
	TR, Toxics, 316(b), NO_x	50-66 GW total (44-60 GW 2020 incremental)	
ICF December 2010	TR, Toxics, Coal Ash, 316(b), NO_x, +CO_2 price	70 GW total by 2018 (including 10 GW of announced retirements)	For Utility Air Toxics Rule, scrubber, ACI, and baghouse assumed for all units. For 316(b), cooling towers on units drawing from coastal and estuarine water bodies. Retirement estimates also reflect cap-and-trade program for CO_2 emissions that begins in 2018.
NERC October 2010	TR, Toxics, 316(b), Coal Ash	10-35 GW by 2018 (excludes 13 GW committed/ announced retirements, which may include non-coal units)	Range reflects 'Moderate' and 'Strict' scenarios. Both assume cooling tower required for 316(b) the primary driver of retirements. For Utility Air Toxics Rule, both assume FGD (with SCR, or ACI + baghouse).

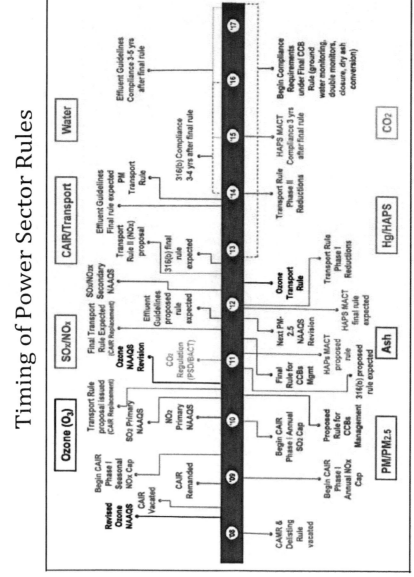

Timing of Power Sector Rules

Source: Edison Electric Institute, http://www.eei.org/whatwedo/PublicPolicyAdvocacy/TFB%20Documents/100525SheaCongressCoalImpacts.pdf (Figure 7).

ACHIEVING A GLOBAL CORPORATE SAFETY CULTURE

Jeff Hannis
Air Products and Chemicals, Inc.

ABSTRACT

Air Products has EH&S Management systems and standards that address the areas of process safety, occupational health, operations, and regulatory compliance. One example system is the Basic Safety Process (BSP). The BSP is a structured and systematic approach implemented by Air Products for the purpose of actively engaging every individual to integrate safety into all levels of the organization.

A company's success depends upon its people. In this paper, Jeff Hannis explains how Air Products and Chemical's safety journey has evolved into achieving a successful safety culture globally - preventing safety-related incidents, fostering behavior based safety, and positively impacting personal safety.

INTRODUCTION

Air Products and Chemicals, Inc. is headquartered in Allentown, Pennsylvania in the USA. The company is a leading supplier of industrial gases, chemicals, equipment, technology, and services to a variety of end user businesses and applications. The company was founded by Leonard Parker Poole in 1940 in Detroit, Michigan. He had the revolutionary idea of producing and selling industrial gases, primarily oxygen, on-site. At the time, most oxygen was sold as a highly compressed gas in cylinders that weighed five times more than the gas product. Air Products proposed building oxygen gas generating facilities adjacent to large volume gas users, piping the gas directly from the generator to the point of use and thereby reducing material handling and distribution costs and risks. A year later, Air Products leased its first oxygen gas generator to a small local steel company and thrived. At the end of World War II, Air Products moved to its current headquarters near Allentown, Pennsylvania, close to the industrial markets in the north-east USA. Today, it also has European headquarters near London, UK and Asian headquarters in Shanghai, China.

THE JOURNEY

Leonard Poole was not only revolutionary in his approach to business; he was revolutionary in regards to employee health and safety as well. Leonard's commitment to establish a safe work environment was so strong that he incorporated this belief into Air Products' Founding Principle (see Figure 1).

Figure 1.0

These words, as one may consider cutting edge in the 1940's regarding 'safety', are still meaningful to Air Products employees today.

The successful evolution of Air Products' Safety Culture not only stems from the robust safety programs that have been implemented over the years, but from another, equally important aspect - 'its people.' Whether our leaders in the company are driving the safety process, demonstrating accountability, or our employees are executing the operational discipline in our plants or on the road, employees are the reason for our safety success across the globe.

Driving success has not come without its challenges as the company experienced a recordable rate close to 3.0 incidents or injuries per 100 employees in the mid-1980's (see Figure 2.0). Leadership in the organization was committed to reducing employee injuries.

Air Products implemented leading safety programs in the coming decades. The company embarked on the Total Safety Philosophy in the 1980's which was similar to Dan Petersen's approach to managing safety by objectives (Dan Peterson - the father of modern safety). In the 1990's, the company expanded a focus on safety with implementation of our Basic Safety Process, a structured approach with forced activities regarding safety behaviors and workplace conditions. In the 2000's, the company worked with Dr. Scott Gellar, a renowned leader in Behavior and People Powered Safety to refine the behavior-based safety aspects of our Basic Safety Process and to drive towards an 'active caring' work environment.

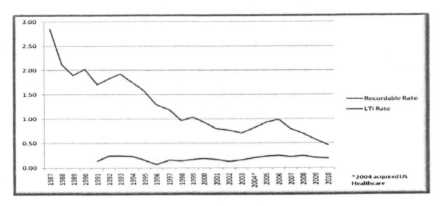

Figure 2.0

Recently, Air Products has reflected on the past 35+ year 'Safety Journey' and formulated a chart based on internal historical data and data from 'The Psychology of Safety Handbook' by Scott Gellar (see Figure 3.0). The data has shown that if our efforts focus just on compliance to regulations (E.g. OSHA Regulations in the USA) that apply to the countries in which we operate, we yield a recordable rate of approximately a 4.0. Furthermore, as our facilities become more structured by executing on the 3 E's, or what we called our Total Safety Philosophy, we are likely to perform at a recordable rate of a 2.5. Through implementation of the Basic Safety Process adopted in the 1990's, data shows we can achieve a rate of 1.0. Today, as we continue to refine our Behavior and People Based Safety Programs, we are heading toward World Class performance and a recordable rate of 0.5.

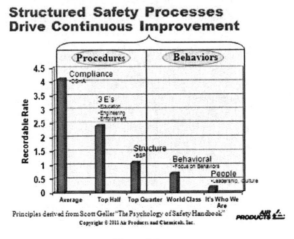

Figure 3.0

KEY ELEMENTS TO GLOBAL SAFETY CULTURE SUCCESS

There are 4 essential elements that drive global safety performance and safety culture improvement at Air Products.

The first, and perhaps most important is safety leadership. John McGlade, CEO and President of the corporation, continues to echo what his predecessors have stated over the decades - 'Safety is a line management responsibility… and nothing is more important than safety; not sales, not profits'. Safety and sustainability continue to be a core value in the company. There are clearly established safety expectations at all levels in the company. Goals and Objectives are cascaded, and the achievement of safety metrics is a key component of annual performance appraisals for every employee. Several 'Key Performance Indicators' (KPI's) and measurements are established for all organizations; including the traditional lagging indicators such as number and rate of recordable incidents and lost time accidents. Establishing and monitoring these KPI's and other indicators help to keep safety as a focus. This has created a safe work environment and is driving safety performance - improving our corporate safety culture.

The second element is our Safety Programs and Standards. In the early 2000's, the company published over 125 EH&S standards that have been applied across the entire company. This was part of our 'One Company' approach in which we combined several legacy standards from our Chemical Operations and our Gases Operations to form standards that apply to each operation globally. Regardless of which facility an employee works, the safety processes and procedures are the same. This establishes consistency in all aspects of the program - including training, execution, auditing, and subsequent improvements.

One specific program that has driven significant improvement in our behavior and conditions based safety performance is the 'Basic Safety Process (BSP).' The program, which applies to each of the 20,000 employees globally, has a number of elements which are required to be executed at all levels in the corporation, and at established frequencies. Execution effectiveness is tracked, and performance measurement scores are communicated up and throughout the organization. A number of the key elements that must be completed are listed below:

- Safety Meetings / Safety Meeting Evaluations
- APT's (Accident Predictive Technique)
- Safety Contacts (individual and group)
- Inspections (planned / leadership)
- Incident Investigation
- Emergency Exercises
- Task Safety Observation
- Job Safety Analysis
- Continuous Improvement Projects
- Critical System Auditing (Safe Work Permits / Confined Space / PPE / etc)

The activities above require different levels of involvement depending on job position and vary from managers, administrators, and operators and drivers. These required activities develop the necessary habits or skills, which in turn build individual and/or team value. Once the employee 'values' the skill or activity, this becomes 'the way we do things' or culture (see Figure 4.0).

Essential to the success of the BSP is documentation. All activities within the program must be documented by the leader, individual, or work teams. This includes all the scorecards used to keep track of the activities. Although this can be a bit onerous, it is essential to the discipline of the process and fosters consistency.

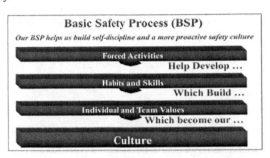

Figure 4.0

Although much of the safety activity an employee experiences each day is driven by BSP, the third key element to building our safety culture is not only 'what' is expected by employees, but 'how' employees participate. This is done in several ways starting with our training programs. Not only do we include the regulatory requirements, but the company provides several other developmental programs targeted to enable our leaders and operations employees succeed. Air Products invests in internal Leadership Education by enrolling select supervisors and managers into programs that teach the necessary soft skills to both manage and lead our organizations successfully in all aspects of the business. Across divisions and at all levels of the company, employees participate in six-sigma and continuous improvement events in which many of the tools used have elements of safety built in. Employees also participate in plant hazard review exercises, providing input to ensure safe operations. Finally, employees are empowered to make 'go', 'no go' decisions regarding safety issues they encounter in their workplace.

The fourth and final element to creating the safety culture at Air Products is Measurement. It is believed that 'what doesn't get measured, doesn't get done.' Air Products has carefully selected critical KPI's that drive performance to not only maintain, but continuously improve our safety performance and culture. The lagging indicators are those we have traditionally measured such as Recordable Rate, Lost Time Rate, Vehicle Accident Frequency Rate, and Process Safety and Environmental Incidents. Where we find more of an impact are those leading KPI's that drive performance such as BSP and sub-element scoring; Incident and Near Miss approval and overdue corrective actions; scoring against our internal electronic site self-auditing program; Mechanical

Integrity completion and overdue items; operating plant hazard review completion and corrective actions; and close out of action items from our Corporate Assurance auditing organization.

CHALLENGES TO IMPROVING / MAINTAINING PERFORMANCE

As a global company with over twenty thousand employees and operating in over 30 countries, there are many challenges in managing our Environmental, Health, and Safety Programs. Although our injury rates have continued to decline over the years, as the company grows in regions such as Asia, there are barriers we must address in order to continuously improve. Air Products' Human Resources organization provides the necessary tools, information, and training for key leaders in the organization to understand the different cultures around the world and how this applies to their leadership skills to properly communicate and influence. Although our EH&S leadership is organized by the regions in which a native language is spoken, the language barrier is a challenge globally for our standards, procedures, policies and training packages and we continue to improve methods to streamline translations and to seek additional ways to ensure that all aspects of safety are interpreted the same by all employees.

As the company continues to grow, we have employed robust integration plans for new acquisitions. A key element to our acquisition integration plans is our safety culture. Not only does this hold true for acquisition, but it is equally important for any new plant start ups.

There are challenges at the facility level as well. Air Products has created tools to evaluate all company sites, and to map out where they are generally positioned in our safety journey (figure 3.0). The purpose of this exercise is to help identify gaps in leadership, programs, or training, so that the gaps can then be closed. Air Products maintains investment in an EH&S group that is staffed by professionals who assist the organization with gap identification and corrective action activities.

There are many aspects to the Air Products and Chemicals Safety Programs that drive good safety performance and our safety culture. The EH&S organization provides the programs, tools, and guidance necessary for our operations to be successfully safe. However, it is the leadership in our organization that continues to value safety and truly places it first and foremost above all else. We believe that a zero-injury work environment is achievable and must be our goal, because --- it's not about the numbers, it's about people.

ABOUT THE AUTHOR:
Jeff Hannis is the Health and Safety Manager for the Americas at Air Products and Chemicals, Inc. in Allentown, PA
Tel: 610 706 2720
Cell: 215 485 7901
Email: Hannisjr@airproducts.com

EMISSION MONITORING IN THE GLASS INDUSTRY

Steve Roosz
Air Tox Environmental Company, Inc.

ABSTRACT

Most glass producing facilities can expect to be required to have Continuous Emission Monitoring Systems (CEMS) at some point in the future. These monitoring requirements will add a new level of complexity and work load to managing the glass furnaces. It-is important to make the correct equipment choice when selecting CEMS to minimize the impact of these added requirements. This paper discusses the requirements a glass plant will face with CEMS, the types of systems available, and offers recommendations for optimizing the CEMS system at a glass plant.

INTRODUCTION

Glass facilities are facing increasing requirements to install emissions monitoring equipment on their furnaces. These requirements are coming from new state rules and EPA enforcement action. States such as California, Pennsylvania, and Wisconsin have passed new regulations in the last few years to limit the amount of NOx emissions specifically from glass furnaces. The EPA has also targeted the glass industry for enforcement action. To date, this has resulted in one consent decree with a major glass manufacturer that requires emission and opacity monitors on all furnaces. Table 1 shows monitoring parameters and the associated analyzers that a glass furnace may be required to implement.

Table 1: Monitoring Parameters and Required Monitors

Monitored Parameter		Required Analyzers							
		Opacity	NOx	SO2	CO	Oxygen	NH3	Moisture	Flow
Opacity	%	X							
NOx	ppm		X						
	lbs		X						X
	tons		X						X
	lbs/ton of glass		X						X
SO2	ppm			X					
	ppm dry			X				X	
	lbs			X					X
	tons			X					X
	lbs/ton of glass			X					X
CO	ppm @ 8% O_2				X	X		X	
	lbs				X				X
	lbs/ton of glass				X				X
NH3	ppmd @ 15% O_2					X	X	X	

CEMS present unique challenges for the glass industry. CEMS add new requirements to plants that are minimally staffed. These requirements are in the form of required checks, maintenance, documentation, and reporting.

The real time monitoring of emissions adds to the work load of the Furnace Manager. Where the furnace emissions may only have been tested once a year in the past, CEMS monitor the data every minute. The Furnace Manager may be faced with new hourly, daily, or monthly limits in addition to annual limits. These present a management challenge because any change in furnace operation and condition can and will affect the emissions.

CEMS may change how the furnace production is limited. Prior to CEMS, furnace limits might be based on pull rate or an emission factor. Once CEMS are installed, the limits will be based on the measured emissions. The limits may be in tons per calendar year, or tons per rolling 12-month period. The furnace production will need to be managed so that the plant stays within these limits. The accuracy of these measurements becomes critical.

There are also unique characteristics of glass exhaust that must be considered in the design of the CEMS. If these characteristics are not addressed in the system design, then the system will have chronic operational problems.

OPERATIONAL REQUIREMENT FOR PLANT WITH CEMS

Most facilities with CEMS need to meet the requirements of 40 CFR Part 60 Appendix B & F, or similar state requirements. In general, these require that the plant perform (and document) the following activities:

Quality Assurance/Quality Control Plan: The site must develop and implement a written Quality Assurance/Quality Control Plan (QA/QC Plan) which describes in detail what periodic checks and preventive maintenance will be done, and what spare parts will be stocked. Some states require that the QA/QC Plan be submitted and even approved.

Daily Checks: The facility will need to perform daily checks as specified in the QA/QC plan. These checks, which must be documented, will include verification that the CEMS passed a daily automated daily calibration check, and other checks to ensure that the system is operating properly. These other checks might include vacuum pressures, temperatures, flows, and other critical CEMS parameters.

Periodic Maintenance: The QA/QC plan will list periodic checks and maintenance activities that are required to ensure the proper operation of the equipment. These might be weekly, monthly, quarterly, semiannual, and/or annual checks, depending on the components and the system design.

Quarterly Cylinder Gas Audits (CGAs): The facility must perform a CGA on each parameter once in a calendar quarter. These must be done 3 times per year, and compare the measured monitor values against certified gas values. Some states require a linearity check, which is similar. A RATA is performed on the fourth quarter as described below

Relative Accuracy Test Audit (RATA): Once a year, the facility must perform a RATA on each parameter. The readings from the facility CEMS are compared to readings from a certified CEMS (usually measured by a stack test company). The state must be notified prior to the testing, and may come on site to observe the RATA.

Corrective Action: The facility must take corrective action whenever the system indicates that there is an emission exceedance, or that there is a CEMS malfunction that is affecting the data quality. The corrective action must be documented and the documents kept in the plant's record system.

Quarterly/Semi-Annual Reports: The facility will need to submit periodic reports on the performance of the CEMS. These may be required quarterly or semiannually, depending on the state requirements and include the amount of time the facility exceeded an emission limit, and the duration of invalid data (when data wasn't available, or when there was a quality problem with the data). The State may also ask for a report showing start and end time, cause, and corrective action for each period of exceedance or invalid data.

Records: All of the information described above must kept by the site for at least five years. In addition, the site needs to keep all emission records for at least 5 years. These requirements are summarized in Table 2.

Table 2: Summary of CEMS Operating Requirements

Requirement	Description	Comment
QA/QC Plan	Written Quality Assurance/Quality Control (QA/QC) plan describes how system will be maintained, and how the plant will ensure that the data collected is valid.	QA/QC Plan may need to be submitted to State.
Daily Checks	Daily checks of system for alarms and to ensure it passed the daily calibration check. Checks will be listed in QA/QC Plan.	Daily checks must be recorded.
Periodic Maintenance	Periodic checks and maintenance activities to ensure the proper operation of the CEMS. These will be listed in the QA/QC Plan.	Periodic Maintenance activities must be documented
Cylinder Gas Audits (CGAs)	CGAs or linearity checks must be done any quarter that a RATA is not performed. The CGA measures how accurately the CEMS measure certified calibration gases.	CGA reports must be submitted to the State.
Relative Accuracy Test Audit (RATA)	RATAs must be conducted at least once per year. The RATA compares the plant CEMS readings against a readings obtained by another set of certified CEMS (typically by a stack test company)	RATA reports must be submitted to the State. Prior notification of a RATA must be given to the State. The State may send someone to observe the RATA.
Reports	The Plant must submit periodic reports that include information on the amount of excess emissions and monitor downtime.	Frequency and details of the reports vary by State.
Record Keeping	All CEMS records must be kept for at least 5 years.	

HOW CEMS CHANGES THE MANAGEMENT OF A FURNACE

Having a CEMS will result in changes in how a furnace is managed. This is the result of continuous measuring of emissions to demonstrate compliance with 1-hr, 1-day, 30-day or longer emission limits. Long term limits (30-day) are easier to manage than short term limits, but it takes longer for a change (or correction) to affect a long term limit.

Any change in pull rate, gas flow, combustion ratio, or batch composition will be immediately reflected in the emissions. This increases the need and urgency of repairing air leaks in the furnace that can cause increases in NOx emissions. Burner system maintenance is also important to control NOx emissions. Raw material contamination can change the redox state of the glass, and result in excess SO_2 emissions, as can weighing accuracy problems with a scale.

If the furnace has a pollution control device, then it will control the impact of these furnace changes. However these will place an increased load on the control device and will result in higher chemical usage, and more control device residue to manage.

Long term furnace emission limits may come in the form of a rolling 12-month sum, in which the total is calculated every month for the last 12 months, or a calendar 12 month sum, which is the total for January 1st to December 31st. These don't move very quickly, but they need to be watched so the plant isn't forced to curtail production to meet the limits.

ELEMENTS OF A CEMS

A CEMS is a system composed of three main elements; the sampling system, the analyzers, and the Data Acquisition and Handling System (DAHS). The sampling system extracts and transports a representative sample of the stack gas to the analyzers. The analyzers determine the concentration of the pollutant in the sample, and the DAHS collect and report the data. These elements are described below.

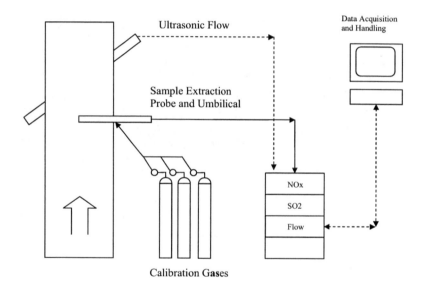

Figure 1: Sample CEMS Layout

Gas Sample Extraction Methods

In most cases, flue gas is extracted from the furnace stack and transported to instruments where the concentrations of the components of interest (NO_x, SO_2, etc.) are measured. Our experience in the glass industry has shown that the method of extraction is one of the most significant design considerations in the CEMS system. It affects the type of monitors that will be used, the equipment that is needed, the amount of maintenance that will be required, and the types of problems that can be expected.

It is essential that the extraction method prevent the formation of water in the sample line. Water can affect the measurement accuracy and can create acid gases which will damage the analyzers. There are three primary types of extraction and delivery methods. Selecting the proper method depends on what components are to be measured and their relative concentrations.

Dry Extraction: Dry Extraction transports the stack gas in a heated line to the analyzers, where the moisture is removed in chillers prior to analysis. The gases then enter the analyzers and are measured in ppm-dry. Dry Extractive systems are common, but have a significant drawback when used on glass furnaces. Glass furnace exhaust can contain SO_3 which forms sulfuric acid mist in the chillers. This acid is difficult to remove, and can cause maintenance problems, increase monitor downtime, and shorten the life of the monitors. This is generally not recommended for glass furnaces.

Dilution Extraction: This approach adds a known quantity of clean dry instrument air to the extracted sample immediately after it is removed from the stack. The diluted mixture has a very low dew point, so the moisture won't condense in the sample line or the analyzers. The sample is measured in the diluted state by the analyzers and the values are corrected for the dilution ratio. The concentrations are measured on a ppm-wet basis which means that it is the volume of the component in a wet sample. By diluting the sample, the problems of acid gases in the sample are avoided. Dilution is usually more cost effective when 3 or less components must be measured in the exhaust.

Hot/Wet Extraction: Hot/Wet Extraction measures the stack gases without removing the moisture. The entire system (sample transport line, analyzer, and pumps) are heated to approximately 350° F to keep any moisture from condensing in the line, and the sample is measured on a hot/wet basis. The concentration is measured as ppm-wet. Some limits are in ppm-dry, and require that the moisture be measured and the sample value corrected. This method is recommended when a multicomponent analyzer is used, or a component must be measured that cannot be measured by dilution. These components include ammonia (NH_3) and hydrochloric acid (HCL).

Gas Monitors

Gas monitors can be single component (most common) or multi-component. Single component analyzers measure one parameter, and a different analyzer is used for each component that needs to be measured. Multi-component monitors use one instrument to monitor a number of parameters. However, they cost more. Since the cost of the base instrument is high, they don't become cost effective until they are used to measure at least 4 components.

Flow Monitors

Stack flow is used to calculate the mass of emissions in units of lbs/hr, lbs/day, etc. Flow monitors determine the velocity of the stack gases in feet/second. The stack diameter and exhaust temperature are used to determine flow rate in standard cubic feet per minute (SCFM). There are a number of ways to measure stack flow. They vary by cost and accuracy.

Calculations using CO_2 Measurements: Some sources calculate flow by measuring CO_2. This is based on natural gas creating a known volume of CO_2 when it burns and this works well in cases where all of the CO_2 comes from natural gas. However, glass furnaces produce a significant amount of CO_2 from the carbonates in the batch. This can lead to large errors in the flow measurement, and is not recommended for glass facilities.

Single Point Flow Measurements: Single point flow monitors measure the flow of the stack at one point. These inexpensive methods will not provide an accurate reading if the stack flow profile changes. There are a number of common furnace adjustments that can affect the flow profile, such as pull changes, educator fan changes, etc. They are also prone to fouling from the particulate in the stack. Types of single point monitors include pitot tubes and hot wire anemometers.

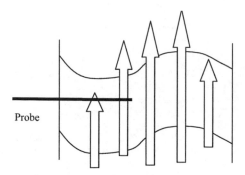

Figure 2: Single Point Flow Monitor Configuration

Through the Stack Monitors: Through the stack monitors measure an average flow across the diameter of the stack. This type includes ultrasonic and optical scintillation monitors. Their readings are less susceptible to changes in flow profile since they measure the flow across the entire width of the flow.

The ultrasonic monitors send an ultrasonic pulse at an angle through the stack to measure flow. The flow causes the ultrasonic pulse to move faster going up the stack (with the flow), and slower going down the stack (against the flow). These time differences are measured and used to calculate a stack velocity. The two heads of the flow monitor must be offset to create the time differences described above. This may require installation of a second platform on the stack, (adding to the costs) or create maintenance issues. They also have temperature limitations of about 650°F and cannot be used on some of the high temperature stacks found in the glass industry.

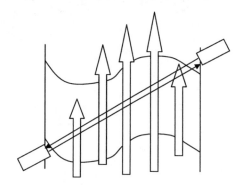

Figure 3: Ultrasonic Flow Monitor Configuration

Optical scintillation monitors are based on the principle that temperature causes turbulence in gases which affects light transmission. This turbulence can be measured and used to determine air velocity. This technology has been used for some time to measure wind shear at airports and has recently become available for stack monitors.

The advantages of OS are that the two heads are mounted straight across from each other on the stack, so no extra platforms are required. And the OS technology works at high temperatures, including those on the hottest glass stacks.

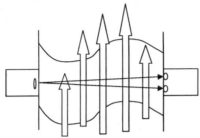

Figure 4: Optical Scintillation Flow Monitor Configuration

DAHS

The Data Acquisition and Handling System (DAHS) or, as it is also known, the Data Acquisition System (DAS), is the heart of the CEMS system. The DAHS consists of a computer, and may include a data logger or PLC. The DAHS is the main Human/Machine Interface with the CEMS.

Typically a DAHS will perform the following functions:

1. Record raw data from the monitors.
2. Use the raw data to calculate values such as flows, emission rates, and averages.
3. Create alarms when there is a monitor problem or an exceedance.
4. Create emission monitoring reports
5. Provide graphs and trends of emissions.

Some highly desirable features in a DAHS are:

1. Simple alarms that tell the operator when they need to take action.
2. Remote access and control of the system via the internet.
3. An email system for alarm notification, and to distribute regular reports.
4. Direct access to the monitor functions and operating parameters. This feature, combined with remote internet access, provides a very powerful remote diagnostics and service tool.

Monitoring instruments use the same EPA approved methods to measure components. However, the DAHS is a software system which is custom configured for each system based on the permit requirements, and any special requirements that the customer requests. There are significant differences in the capabilities, ease of use, and look and feel of different systems, and there are a number of systems available. Special attention should be given to evaluating and selecting a DAHS!

ELEMENTS OF A SUCCESSFUL CEMS FOR THE GLASS INDUSTRY

Now that the components of a CEMS system have been discussed, it is time to describe how they should come together to meet the unique needs of the glass industry. The goal of any monitoring program is to provide maximum data capture (system uptime), provide the tools the operator needs to stay in compliance, and to do so with a minimal workload to the plant operating personnel.

Daily Checks: The system should be designed so that all of the daily checks are done at the DAHS in the control room. The QA/QC Plan lists the daily checks that are needed to ensure that the system is operating properly. These may include sample flow rates, system pressures, vacuum pressures, calibrations and calibration gas pressures, and system alarms.

Many systems require the operator to leave their workstation and perform these checks on the instruments themselves. The CEMS may be located some distance from the control room, which means the operator must stop monitoring the furnaces to do the checks.

A better approach is to put sensors on all the critical parameters, and tie them into the DAHS. This allows all required daily checks to be done in the control room without the operator leaving their station. This simplifies the checks, helps to ensure they get done, and tells the operator when there is a problem that needs attention.

Alarms: Alarms need to be simply displayed, not subject to interpretation, and should clearly tell the operator that action is needed immediately.

Alarm Response: Alarm responses should also be very simple. Furnace operators are trained to operate furnaces, not CEMS. They should be expected to respond to emissions problems by managing the furnace. There should be an outside service available for managing CEMS hardware or software problems.

The outside service should include a 24/7 service number, an email alert system, and an internet connection to the system. The internet connection allows the service provider to immediately connect to the system and start troubleshooting. If the system has been designed with adequate connectivity, the service provider can perform a detailed diagnosis and often fix the problem remotely. If the problem can't be corrected on-line, then the service provider can determine if the problem can wait until the next scheduled maintenance, or if a technician needs to be sent to the site. In all cases, decisions about the CEMS are being made with the help of a trained CEMS professional.

Periodic Maintenance: Periodic preventive maintenance (PM) should be required no more often than quarterly. Some CEMS systems require monthly or weekly checks and PMs, which need to be done by plant personnel. This requires that the plant personnel be trained to do the PM checks, and to document them. Failure to perform or document a PM may result in a reportable deviation.

A better way is to design the system so that PMs are only done once a quarter. Most sites will have a CEMS service vendor on-site once per quarter to do the CGAs and the RATAs described above. While on-site, they can also do the required equipment PMs. This ensures that the PMs are done by trained technicians, and minimizes the work load for the plants.

Spare Parts: The plant must have spare parts available to repair the CEMS. The equipment manufacturer determines the required spares and quantities, and they are listed in the site's QA/QC plan. Most plants buy a set of spares when they purchase the CEMS, and keep them in their general store room. The spares can be quite expensive, sometimes costing $20,000 or more. And many of the expensive spares are delicate electronic components (boards, power supplies, sensors, etc.) that can be damaged if not handled and stored properly.

An alternative is for the site to only keep the consumable spares on-site (parts that are replaced during the quarterly PMs), and to have the CEMS vendor stock the other spares. The vendor must agree to 24-hr delivery of the parts to prevent excessive downtime on the monitors. The vendor ensures that the parts are properly stored and maintained, and the plant is only charged for the parts when they are used.

Table 3: Summary of Desirable Features in a CEMS System

Key Point	Description	Benefit
Daily Checks	Instrument equipment so that daily checks can be done either remotely or in the furnace control room	Operator does not need to leave work area to perform checks; checks can be done by third party
Alarms	Unambiguous alarms that indicate that action needs to be taken.	No guessing what an alarm or warning means.
Periodic Maintenance	No planned maintenance more often than quarterly	All maintenance done by service technicians during quarterly service visits.
Spare Parts	Agreement with vendor to supply spare parts within 24 hrs.	Plant does not need to stock expensive and delicate spare parts
Email capability	System emails selected alarms to a predetermined distribution list	Plant personnel immediately made aware of CEMS problems
Remote access to system via internet	Service personnel can perform troubleshooting and take corrective action without being on site. Plant personnel can remotely access the system to run reports.	Improves response to problems, reduces downtime, and avoids cost of on-site service visits.

SUMMARY

Most glass producing facilities can expect to be required to have CEMS at some point in their future. These monitoring requirements will add a new level of complexity and work load to managing the glass furnaces which include performing and documenting periodic checks and maintenance on the system, and responding to the CEMS information to keep the furnace within its permit limits. There are a number of options in CEMS. Some of these options work well in glass, while others will create problems for the site- either by added maintenance, or by potentially overstating emissions. The hardware, software, and options that a plant chooses when installing a CEMS system will have a lasting impact on how well the system performs, and how much extra work will be required to meet the new requirements.

Recycling and Batch Wetting

DESIGN OF A NEW 25 TON PER HOUR WASTE GLASS PROCESSING PLANT FOR RUMPKE

Christian Makari and Stefan Ebner
Binder+Co AG, Austria

ABSTRACT

In late 2011 Rumpke Recycling, based in Dayton, Ohio will begin operating one of the most modern waste glass processing plants in the world. Binder+Co AG, an Austrian company and a market leader in technologies for waste glass processing, designed a tailor-made plant for the specific needs of Rumpke Recycling and its customers. Sophisticated state-of-the-art machinery will be used to separate contaminants from a raw feedstock to produce two valuable products – glass cullet, colour separated and cleaned for the glass bottle manufacturing industry, and high quality glass powder used by the fibreglass insulation manufacturing industry.

The new glass processing plant will have a capacity to process up to 25 tons per hour of residue material created in waste treatment plants (MRF) in the Dayton area. As this material contains a significant fraction of non-glass content, efficient systems for pre-processing the waste product are essential for accurate and precise separation of the contaminants. Conditioning technology such as screening systems and crushers, successfully adapted by Binder+Co for the mining industry, and other proprietary Binder+Co technology such as organic separators and suction plants, specially produced for the recycling industry are used to efficiently extract non-glass waste. Drying is accomplished in DRYON, a proprietary fluidized bed drying system. CLARITY optical sorting technology creates colour-separated streams of both amber and flint glass for the glass container industry. Separated colours pass through additional optical separators to remove ceramic, refine colour purity and create a finished product. Finally, the remaining non-colour-separated glass as well as finer fractions are recombined and processed to create a finely ground -12 mesh product for the fibreglass insulation industry. The upstream processing in the Binder+Co plant helps ensure that the glass powder has a minimum content of organic contaminants, thus meeting a key need of this product and helping to produce environmental friendly insulation materials. The new recycling plant of Rumpke will start up in October 2011, providing a new milestone for sustainable environmental technology.

The paper provides a technical description of this new glass processing plant. It describes all the steps from characterization of the incoming feedstock to definition of the final product quality, including all the necessary cleaning and separation steps leading to the two desired final products: -plus 1/2" glass cullet and fine grind material. The produced final products will be delivered to the two customers that have been involved in the project from the start, Owens Illinois from the glass container manufacturing industry, and Johns Manville, from the fibreglass insulation manufacturing industry.

PROCESS INPUT MATERIAL
Figure 1 shows an example of the composition of the incoming feedstock. This picture shows that the input material contains a very high amount of non-glass contaminants. This non-glass fraction can have a quantity of up to 15 % in weight and mainly consists of following materials:

- CSP (ceramics, stones and porcelain)
- Plastic films and foils
- Whole plastic bottles
- Plastic caps and corks
- Shredded paper
- Food product waste
- Ferrous and non-ferrous material
- other

These contaminants must be removed in the initial processing stages to ensure properly working optical separation processes and product meeting the required final product specifications. Extraction of non-glass particles is most important in a pre-treatment stage to avoid unfavorable irritation or blinding of the optical detection process. Food waste has a sticky character initially but loses this property after being dried, and does not influence later processing steps.

Figure 1: Input Feedstock

Another obstacle in processing glass cullet coming from a Materials Recycling Facility or MRF, is the moisture content. Since the feedstock material is stored outside, and therefore exposed to weather conditions, it can be wet one day and dry the next day, depending upon weather. To ensure a stable and steady production process, and also to meet specifications for the fiberglass industry, a drying step must be incorporated into the process design.

FINAL PRODUCT QUALITIES
The waste glass processing plant designed by Binder+Co AG will process waste MRF glass material and produce two different high-quality, low-contaminant glass products for use as raw materials in glass-making operations in the container glass and fiberglass industries.

Two main products will be produced:

- Glass cullet (flint and amber)
- Fine ground mixed glass

Glass cullet qualities:

- Flint   Purity by color: > 98,5 %
  CSP content: < 50 ppm

- Amber   Purity by color: > 98,5 %
  CSP content: < 50 ppm

Fine grinding material qualities:

LOI: < 0,25 %

To meet the maximum allowable LOI (loss on ignition), the organic content (typically approximately 15 % of the actual input material) will be reduced by about 98-99%.

PLANT AND PROCESS DESCRIPTION
The Rumpke waste glass processing plant can be divided into four main areas: Organic Removal, Drying, Optical Separation, and Fine Grinding. Figure 2 shows a schematic of the entire system with these areas highlighted. Within each area there are multiple sub-processes.

PRE-PROCESSING AND ORGANIC REMOVAL STAGE
Due to the high amount of non-glass material, especially organic material, contained in the feedstock, the first processing stage is equipped with very specialized machinery from Binder+Co to aid in removing organics while minimizing loss of glass. Figure 3 shows the first part of this organic contaminant removal stage.

Input material is fed from the **infeed hopper (1)** via the **discharge feeder (2)** onto a **belt conveyor (3)** at a rate of approximately 25 tons/hour. On this belt conveyor the material weight is detected with the **belt weighing system (4)**. According to the calculated material bulk density the belt weighting system automatically regulates the speed of the belt conveyor (3). Ferrous material is removed by the **overbelt magnet (5).**

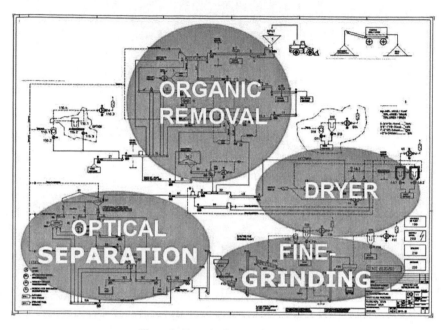

Figure 2 : Rumpke Process Stages

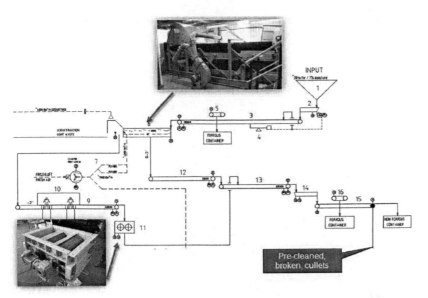

Figure 3: Initial Organic Removal

The **Resonance screening machine (6, photo)** is the first main screening step in this process. It is equipped with a finger screen in the upper deck and screening mats in the lower deck. The oversize fraction, consisting of cubical and voluminous contamination, partly broken glass bottles (bottle necks combined with corks, caps and rings as well as bottle bottoms) is conveyed over the finger screen elements in the upper deck. The Resonance screen is equipped with a **blowing unit (7)** that here extracts light, dry and relatively large organic particles (e.g. plastic bottles) from the main input stream. The remaining material larger than 2 inches goes onto the **picking belt conveyor (9)**, where remaining bulky contaminant items are removed by a employee at the **picking station (10)**.

In a second screening step the 0 inch to 2 inch fraction is effectively screened off with flexible screening mats in the lower deck of the screening machine. This fraction then goes directly to **belt conveyor (12)**.

The size fraction larger than 2 inches is fed to the **double roll crusher (11)**. This process provides crushing glass material into grain size range below 2". A double roll crusher provides a smooth crushing process to prevent the production of too much fine material, enhancing the effectiveness of the optical sorting processes downstream.

The material streams from **belt conveyor (12)** and the **double roll crusher (11, photo)** are recombined on **belt conveyor (13)**. Here an **eddy current (15)** removes non-ferrous metals from the material stream. These non-ferrous metals are collected in a separate container. A second **overbelt magnet (16)** also removes ferrous metals, which are collected separately from the non-ferrous contaminants.

Organic removal continues and is shown in Figure 4. The **screening and distribution feeder (19)** continues the organic removal process. Similar to the Resonance screen, this screen is also equipped with a **blowing unit (20)** to remove additional light organic material. To increase the efficiency of this processing step all material smaller than 3/8 inch is first screened off in (19). This fine fraction contains finely sized organic material which will be removed on **vibratory feeder (24)** which is equipped with a special suction system.

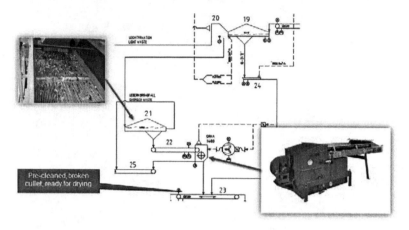

Figure 4: Additional Organic Removal and Size Reduction

After passing the **blowing unit (20)** the coarse (>3/8 inch size) fraction moves on to **screening vibratory feeder (21, photo).** This screening vibratory feeder is equipped with a finger screen (or bar screen) to once more remove organic items, especially still voluminous (after crushing) items like plastic caps, corks etc. The undersize from item 21 will in a further step be processed by the **ORKA organic separator (red machine shown in small photo).** The organic separator is a very specialized piece of equipment, for a last organic removal step, before the pre-cleaned cullet stream will go to the drying stage. Design of this machine is proprietary and is not described.

DRYON DRYING STAGE
One main purpose of the drying stage is to have year-round stable sorting efficiency on the optical separation stage trough, regardless of weather variation and moisture content of the input feedstock. Also, though container manufacturing specifications do not require 100% dry material, the need to remove most of the LOI in the fibreglass cullet product requires a very low moisture content. In case the feedstock material is dry and does not need to be additionally dried, the drying stage can be bypassed via **reversible belt conveyor (23)** (see Figure 4).

The drying process is shown in Figure 5. The Rumpke plant is equipped with DRYON, a **fluidized bed dryer (41)** by Binder+Co AG. This kind of drier is especially designed for waste glass processing. The glass cullet is fluidized in the dryer by hot air streaming through a perforated plate.

The design minimizes direct contact between abrasive cullet and the drier, thus minimizing drier wear and maximizing life of the drier. The smooth conveying method within the drier also minimizes production of inadvertent cullet fines. The drier spreads the cullet out into a thin single layer, enhancing contact between hot air and glass cullet, thus leading to higher drying efficiency. The drying zone where adherent water is evaporated uses approximately 80% of the length of the drier. The drying temperature is designed to always be less than 572 °F to assure a safe drying process.

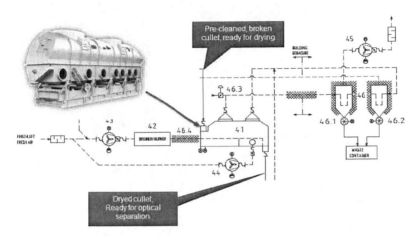

Figure 5: The Drying Process

The remaining 20% of the drier length is designed as a cooling stage. Here the fluidized bed is generated by the use of ambient air. While cooling the glass cullet the cooling air is heated and subsequently used for preheating burner combustion air. This use of preheated combustion air in the drying process provides a significant system advantage, keeping the process "green" by minimizing energy consumption.

Note that this type of dryer will also remove organics to a certain extent. The dried fine organics are extracted by the exhaust air and separated in the air filter system.

CLARITY OPTICAL SEPARATION STAGE
After the cullet stream is processed, sized, cleaned, and conditioned in the first two stages, it is ready for optical color separation by CLARITY. This highly specialized machinery is a latest state of the art optical separator. It can be equipped as a two- or three way system (see Figure 6).

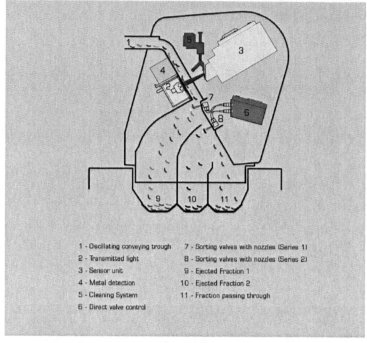

1 - Oscillating conveying trough   7 - Sorting valves with nozzles (Series 1)
2 - Transmitted light   8 - Sorting valves with nozzles (Series 2)
3 - Sensor unit   9 - Ejected Fraction 1
4 - Metal detection   10 - Ejected Fraction 2
5 - Cleaning System   11 - Fraction passing through
6 - Direct valve control

Figure 6: Clarity Optical Sorting

The optical separation stage creates two products, which will be high quality input raw materials for the container glass industry:

- Flint cullet
- Amber cullet

The optical sorting system is charged with cullet having a size range from 1/8 to 2".The material stream runs over a glass chute at a 60° inclination. The cullet is arranged in single pieces by means of acceleration.

The described sorting unit is a transmission system, where that cullet on the detection chute is transilluminated using light of a special color temperature. The transmission information is recorded by highly sensitive CCD cameras and analyzed. The valves mounted on the blow-off strips are activated with precise timing based on material flow rates to remove only the contaminant or glass color desired. Contaminants and predefined colors are blown off into the appropriate tracks according to the programmed menu. The sorting paths are freely selectable.

The coordination between camera units and lightning is carried out automatically. The system is equipped with differing numbers of valves depending on the grain size of the material to be processed. The valve opening times are automatically adjusted to the size of the object. This guarantees the highest possible sorting quality with the lowest loss of glass. This optical separator has been designed in module form to enable a trouble free integration in existing glass recycling systems. All these units can also be upgraded with sensors for detection of heat resistant glass or glass with a lead content.

The Rumpke waste glass processing plant optical sorting stage is equipped with three units (see Figure 7, photo). On the **screening and distribution feeder (83)** cullet smaller than 1/2 inch is screened off and transported on a **distribution feeder (90).** This feeder is equipped with special suction hoods for the removal of dry organic material. After this suctioning the material goes directly to the fine grinding stage.

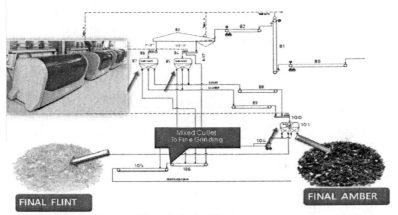

Figure 7: Optical Sorting

The screening and distribution feeder (83) divides the remaining cullet into two grain size fractions, ½ inch to 1 inch size and 1" to 2" size. In the first separation step two optical sorters work in parallel. One unit handles the smaller grain size fraction and the other unit handles the larger size fraction. Splitting the stream into two fractions improves the accuracy of the optical separation.

The cullet streams are fed to the optical separators by **distribution feeders (84 and 86).** The purpose of these feeders is to assure a single layer of cullet arriving to the optical separators. These feeders are also equipped with special suction hoods for the extraction of additional dry organic material.

Both optical separators (85 and 87) are equipped as 3-way machines. Each ejected stream of cullet is blown into a separate chute of "way". Flint cullet is ejected in the first way and transported to the second stage via a chute. Amber cullet is ejected on optical separators 85 and 87, into the second ejection way, and also transported to the second separation stage in a separate chute.

The non-separated cullet fractions or throughput material, mainly green glass cullet, is transported directly to the fine grinding stage.

Separation stage two is equipped with a **optical separator in a 2-way design (101)**. The purpose of this stage is to refine **flint and amber cullet to produce a furnace ready product**. This second stage (101) is designed as a divided unit, which means, it processes flint on one half of the machine and amber on the second. The ejected material is also reused, since it goes to the fine grinding plant.

FINE GRINDING STAGE

All the fines fractions from previous stages, plus the throughput from the first stage of optical separation, and the ejected material from the second optical separation stage go to the fine grinding plant, shown in Figure 7. This inventive concept design where the cullet is automatically separated and processed to two different sized final products not only greatly minimizes glass losses but also minimizes the rehandling and reprocessing of glass typically required to make the fine grind product.

Intermediate hopper (F11) (see Figure 8) provides a steady feed to the **vertical impact crusher (F13)**. The crushed material gets final sizing in the **screening machine (F15)**. This screen is also equipped with multiple suction points for a further reduction of organic material. The oversize from the screen returns to the intermediate hopper to rerun through the crushing process.

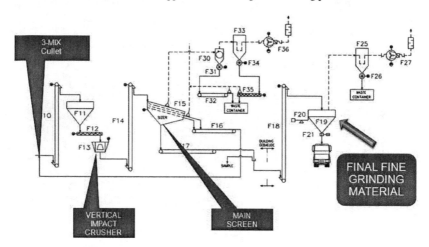

Figure 8: Fine Grinding Plant

The final product is a finely sized material with low organic content, and low metals and CSP content, providing a high quality input material to the fiberglass industry.

Fig. 9 shows some of the advantages of Rumpke's new waste glass processing plant in comparison with the old plant. Figure 10 shows the technical data sheet for the new Rumpke plant.

Comparison of Rumpke´s old Plant and the new Waste Glass Processing Plant	
old Plant	new Plant
Only production of fine grinding material	Flexible production of two valuable products for two customers
Organic removal not satisfying	Highly advanced pre-processing and organic removal stage
High energy consuming drum dryer	Efficient fluidized bed dryer system DRYON
Labour intensive handpicking	Highly effective automized color separation by CLARITY
Maintenance intensive	Maintenance friendly through advanced plant design

Figure 9: Comparison of Rumpke Old and New Plant

Technical Data Sheet Rumpke Recycling Waste Glass Processing Plant	
Start up date:	October 2011
Input Capacity:	25 t/h
Electrical power consumption:	400 kW
Suction air supply:	18.800 cfm
Footprint (l x w x h):	162' x 74' x 19'

Figure 10: Technical Data Sheet – Rumpke Recycling Waste Glass Process in Plant

SUMMARY
The Rumpke Recycling waste glass processing plant is an example of an innovative design incorporating new cullet processing concepts and state-of-the-art technology and equipment. It more fully automates the processing of cullet products for two completely different glass industries, reduces rehandling of cullet, minimizes process losses, and decreases energy by use of preheated drier combustion and highly efficient fluid-bed drying. The old Rumpke plant was not capable of processing all of the glass collected in Rumpke businesses and was not capable of processing to meet Glass Container Manufacturing needs, The new plant can produce not only glass cullet for glass container manufacturing customers but also for fiberglass manufacturing processes. The increased system capacity will also be able to take care of Rumpke's processing needs now and in the future.

GLASS RECYCLING TECHNOLOGY OF TODAY

Hoser Moser

ZIPPE Industrieanlagen GmbH, Germany

GLASS RECYCLING TECHNOLOGIES OF TODAY

The skilled application of raw materials and energy as well as the intelligent recirculation of both is becoming more and more important in our industrialized world. This is valid in particular for the glass industry, being a branch with high energy and raw material consumption.

The gains are significant: Apart from saving good money, coming from lower glass melting costs, and preserving raw material sources, a valuable contribution to environmental protection can be achieved. Both targets are well accepted by our society and often supported by politics.

Recycling in the glass industry is a wide field with regards to the great number of existing glass branches and products. Tailor made solutions had to be developed to work reliably day and night with lowest wear and maintenance possible.

For all of them, however, the basic demand of recycling is valid: The waste glass occurring during production in form of hot and cold glass as well as recollected glass products has to be recycled and processed to achieve a cullet material suitable for remelting. Cullet has to be crushed to a predicted grain size and have to be free of hazardous contaminations affecting the furnaces and the quality of production.

CONTAINER GLASS
During production in the container glass industry factory waste glass is occurring at the hot and cold end in form of hot formed gobs and hot articles in the IS-machines area as well as cold from the inspection line.

Hot waste is charged in water-filled scraping conveyors to be collected, granulated and transported within the recycling loop.

A good granulation is depending on a sufficient long dwell time of the hot glass in the cooling water. That cooling water, by the way, can be recycled in systems removing fine glass particles and oil, cooling the water and reintroducing it into the cooling circuit.

An efficient wear lining of the scraping conveyor tank floor and the scraper bar edges or even tank execution in stainless steel are essential to reduce corrosion and wear and to minimize the input of metal abrasion into the cullet material.

Floor lining by basalt tiles or cast metal plates are common use, but also more and more the forming of a bed of granulated glass covering the tank floor makes it possible that glass is transported on glass which is resulting in low wear.

Scraper conveyors are available in various lengths and executions. They find their application in all kind of glass production or even for furnace tapping. The longest one ever supplied worked in Brazil for container glass and had a length of 54 meters.

Some years ago in the United States, a well-known container glass factory operating 5 furnaces with a total capacity of abt. 2,000 ton of glass per day decided to install scraping conveyors for processing their factory hot waste glass. The granulated glass has to be combined with returning cold end waste and crushed before being recharged into the melting loop.

Crushing by hammer mills are giving excellent results, because the grain size of the produced cullet can be influenced by various adjustments, so as speed, number of installed hammers and variation of distance between rotating hammers and crusher bar. Compared to other crusher types, in particular to jaw crushers, hammer mills having a much lower power input and produce a much more uniform grain size.

ECOLOGY GLASS

Post-consumer glass, also called ecology glass, mainly used for producing container glass, is playing a more and more important role. Melting glass with
90 % and more of cullet is not unusual nowadays. Each 10 % of ecology glass input into the furnace reduces the melting energy by 2,5 - 3 % - and it is interesting to know that one ton of ecology glass replaces 700 kg of sand, 190 kg of soda ash, 150 kg of limestone, 80 kg of dolomite and 50 kg of feldspar.

And – last, but not least: Each ton of ecology glass reused for melting avoids 670 kg climate damaging CO_2 (EU average). This means that using 100 kg ecology glass instead of 100 kg raw materials, the CO_2 emissions can be lowered by 60 %.

Ecology glass recycling has to concentrate on reliable detecting and removing all material contaminating that glass, so as ferrous and non-ferrous particles, stones, ceramics, porcelain, vision ware, glass ceramics, etc. to produce a final cullet grain size of max. 30 – 40 mm without creating too many fines. Only 12 % of it should be below 5 mm in grain size.

The prescriptions for treated ecology glass are very stringent. So, there is allowed that one ton of glass may only contain:

25 – 35 g	CSP (ceramic-stones-porcelain)
25 g	glass ceramics
5 g	ferrous metal
5 g	non-ferrous metal
1 g	lead
5 g	aluminium
200-500 g	organics
60 g	plastic materials
1.500 g	paper, cork, wood
100 g	opal glass
Moisture:	2 – 3 %

And less than 12 % material under 5 mm

Efficient plants having a throughput of up to 40 tons/hour mainly used by commercial cullet dealers are impressive and sophisticated installations.

Manual work for removing big size contaminations out of the incoming material is combined with High-Tech involved in the controlled separators with special cameras for organics (in particular ceramic, stones and porcelain) and all metals, say ferrous and non-ferrous ones.

Colour glass separation is a topic today as well.

We think however that only the separation of about 3 – 4 % of non dominating colour out of flint glass is economically recommendable. We call that colour improvement.

Strict limits of impurity quantities still left in treated ecology glass have to be respected to avoid that loads of cullet will be returned to the dealers, because of refused acceptance.
Grinding cullet and impurities (CSP) under a grain size of 1 mm (for dissolving CSP) is not very common anymore.

FLOAT GLASS / FLAT GLASS

When it comes to recycling of float glass waste developed during production, inline crushing and collection systems are required. Cut off edges and rims as well as faulty sheets have to be broken without affecting the continuous drawn glass ribbon conveyed on the roller track.

Float glass edges are falling into steel plate hoppers located beneath that roller track. Adapted crushers at the exit of those hoppers break then to pieces.
Float glass sheets are directed into plate glass crushers by tilting roller track sections. The produced cullet are transferred into hoppers situated below.

Producing those hoppers in double steel plate execution installing a rubber layer between both walls will make a good contribution to noise reduction. Both cullet fractions do not yet have the required grain size for remelting, but are only products of a precrushing step having the task to produce a transportable size of glass parts. Collection is made by using belt conveyors and a final breaking is carried out in a hammer mill crusher.

It is compulsory that all crushers are equipped with efficient dust collection systems to avoid that glass dust will come in contact with the glass ribbon. No need to say that only the best wear resistant materials and hard-facings available are used for the crusher tools. This is necessary in particular when it concerns solar glass or thin flat glass recycling.

Here it is an absolute "must" to keep the iron input produced by crusher tools really as low as possible – a request, by the way, being valid for all equipment processing unit handling melting materials.

VERY THIN FLAT GLASS

Very thin flat glass (TFT, LED, display glass or others) being a flexible material which does not give great mechanical resistance against crushing tools is a challenge for recycling operations.

The glass has to be broken without creating any iron input. We have developed a pilot system working with a pair of rotating shafts equipped with special plastic cams having a suitable form for inline breaking, better cracking the thin glass ribbons.

Produced cullet is falling into an exit hopper being lined with natural rubber. So, there is no contact with iron during the entire process. We could imagine that this is a way to go and will continue our developments.

Recycling of safety laminated glass is possible. This applies for windscreens for cars up to bullet proof glass. The sheets or sheet packs will be precrushed in a roller crusher to achieve fragments of abt. 4 by 4 inches. Those will be charged into special crushers where glass will be broken and removed from the plastic foil up.

The vibratory screen is finally separating glass cullet and foil fragments and an installed magnetic separator removes all tramp iron particles.

FIBER GLASS

Recycling of fiber glass in form of glass wool and continuous filament is increasing in its importance as a consequence of growing production quantities.

Apart from the problem of having waste in a rather inconvenient form, say big fluffy volumes of low weight waste or rolls of insulation mats, etc., the fibers are also coated with toxic binders.

Those binders or coatings have to be removed prior to remelting, because they would cause quality problems at production and harm the environment.

Our original idea had been to crush the fiber waste mechanically and to vaporize the binders in a heated rotary kiln. Vaporized binders would then have to be burnt in high temperature chambers resulting in destroying the hazardous toxic contents.

The obtained pure glass would have to be crushed and mixed after with the batch, say treated as a raw material. Unfortunately this system was not considered as being economically efficient enough.

Today in big fiber glass factories the fiber waste is mechanically crushed and melted in special ovens being designed just for this purpose. The high temperature combustion gases are burning the binder and are cracking the toxics. After filtering, the cleaned gases are released. The liquid glass leaving the oven is granulated in a water bath for being reused either for remelting or for other purposes.

SUMMARY

All over we can say that recycling is an absolute must today with regards to economical and environmental background. Cullet is a classic example of a win-win situation. Existing technologies have to be steadily improved and new ones developed to meet the challenges of recycling new glass products coming up. All those efforts are not cheap, however, are binding development and manpower resources and might not always meet the expectations – but they are a good investment for a better and cleaner world of tomorrow.

CORD TESTING USING THERMAL SHOCK: VIRTUE OR VICE?

Gary L. Smay and Henry M. Dimmick, Jr.
American Glass Research

ABSTRACT

Abraded thermal shock tests have been used for decades to justify decisions relative to packing ware that contain certain levels of cord streaks. This paper explores the science behind these tests and indicates that the situation is more complicated than originally thought. In order to take full advantage of this type of test, glass thicknesses must be taken into account. Otherwise, the results of these tests are meaningless and potentially misleading.

INTRODUCTION

Cord is defined as a streak of glass whose composition differs from the bulk glass (1). Due to these compositional differences, certain properties such as thermal expansion, viscosity and density are affected (2). The most important of these is the coefficient of expansion as it controls the generation of stresses in both the cord streak and the surrounding glass.

Depending on the composition of the cord streak, the thermal expansion of the cord will either be less than or greater than the surrounding glass. For cord with a thermal expansion greater than the glass, a tensile stress will be generated in the cord streak. Since glass only fails under the action of tensile stresses, the magnitude of the stress associated with this cord must be analyzed properly and the source of the cord must be identified and corrected. Cord streaks exhibiting compressive stresses are only important relative to the magnitude of tensile stresses that are induced in the glass surrounding the cord streak. As with tensile cords, the magnitude of the induced stress must be analyzed properly.

These tensile stresses, either induced in the surrounding glass or generated directly in the cord streak, can adversely affect the performance of containers. The presence of tensile stresses associated with cord streaks mathematically adds with tensile stresses created in the glass by loads applied to the container. This sum can be greater than that which is normally expected due to the applied load, bottle design and glass thickness distribution. Thus, cordy bottles have the potential to fail at loads that are far below those that the container is typically expected to endure.

The onset of a cord problem can originate from a wide variety of sources relative to batch and furnace operations as summarized in Table I. The onset of cord can be relatively rapid; however, once the cord occurs, it can exhibit great longevity as depicted in Figure 1. Thus, there is always a conundrum as to when containers that initially have excessive levels of cord stress can once again be properly packed and the production floor is returned to normal operations.

Historically, this problem has generated a great deal of debate among plant personnel with competing interests between maintaining a high percent pack and assuring the containers are of proper quality. A methodology that has been routinely used to address this debate is the use of abraded thermal shock tests. The assumption has always been made that if containers afflicted with tensile cord streaks pass the abraded thermal shock test, then they are suitable for packing. This test procedure and its appropriateness with regard to these questions will be critically examined in the current paper.

221

CORD DETECTION AND MEASUREMENT

Cord streaks can on occasion be detected by use of a polariscope. When bottles containing cord streaks are viewed in such a device, the cord will sometimes exhibit either blue or orange colored streaks. However, some detection limitations exist relative to the positioning of the cord streaks in comparison to the set-up of the polariscope and with regard to the intensity of the cord stresses. Due to these limitations, polariscopic examination should not be relied upon as a sole indicator of either the presence or absence of cord. It is far more reliable to use routine examination of ring sections as a means to detect and measure the stresses associated with cord streaks (3). In this method, the colors observed in the cord streak are the result of the interaction of polarized light with the cord stresses creating a certain amount of optical retardation which can be accurately measured with a suitable compensator. This retardation, R in nm, is converted into stress, S in units of psi, by use of the following relationship:

$$S = 2.2\,R/t \quad (1)$$

where t is the optical path length, in units of inches.

REACTION TO CORD – CREATION OF INDUSTRY GUIDELINES

The establishment of industry cord guidelines can be traced to a paper published in 1939 (4). In this paper, glass quality was judged by a letter designation ranging from A to E, according to the amount and intensity of cord that was detected in a ring section. Based on empirical observations, cord grades of A – C were judged to be non problematic while grades of D and E potentially could adversely affect container performance. The concept of a cord stress index was also introduced in this paper. According to this concept, the stresses calculated by equation 1 should be weighted by the position of the cord streak relative to the original outside and inside glass surfaces. These weighting factors take into account that tensile cord on the outside glass surface has the greatest potential to adversely affect container performance while cord on the inside glass surface is less important and cord that is buried in the interior of the glass has essentially no effect on performance.

In addition to glass quality letter grades, approximate guidelines for the maximum tolerable cord stress for both refillable and non-refillable containers have been established (5). These guidelines were based on the performance of containers with various levels of cord and were established to assure that cord stresses would not become excessive to the point where containers would fail at unusually low forces. These suggested guidelines have been adapted today by individual glass container manufacturing companies to meet their specific management philosophies and customer tolerance levels.

GLASS FRACTURE CRITERIA

For any glass object, fracture occurs when:

$$\text{Tensile stress} \geq \text{Surface Strength} \quad (2)$$

In the case of glass containers containing some level of cord stress and being subjected to thermal shock, equation 2, can be expanded to:

$$\sigma_{ts} + \sigma_{cord} = S \quad (3)$$

where S is the glass surface strength and σ_{ts} and σ_{cord} are the stresses associated with thermal shock and cord, respectively.

Tensile stresses generated by thermal shock are given by:

$$\sigma_{ts} = X(\Delta T) \quad (4)$$

where ΔT, is the temperature differential, in degrees F, and X is the stress index for thermal shock. Substituting equation 4 into equation 3 gives:

$$X(\Delta T) + \sigma_{cord} = S \quad (5)$$

This equation can then be rearranged to give σ_{cord}:

$$\sigma_{cord} = S - X(\ T) \quad (6)$$

where, X is dependent on glass thickness as shown in Figure 2. Equation 6 was used in the current study to determine the magnitude of tensile cord stress that can be detected by abraded thermal shock tests for various abrasion severities and thermal shock temperature differentials.

ABRADED THERMAL SHOCK CONSIDERATIONS

As discussed earlier, a highly stressed tensile cord would have very little, if any, effect on performance if it is buried in the interior of the glass. However, if that same highly stressed cord was present on the outside glass surface, it could seriously affect the performance of the container potentially causing it to fail at abnormally low applied load levels.

Thus, once a cord problem has occurred, it is important to assess the magnitude of stress associated with the cord and to determine if the cord is present on the outside surface of glass containers. Results of these determinations are used to make decisions regarding whether the ware should be discarded and, if discarded, when it would be appropriate to begin packing the ware again. One technique to accomplish this objective would be to frequently analyze ring sections that are cut from various sidewall heights from the afflicted bottles. However, this process is time consuming and glass companies have sought to find an alternative method to detect the presence of highly stressed cord on the outside glass surface. The abraded thermal shock test was devised in response to this concern. Not only was this test intended to detect the presence of a highly stressed cord, it was also intended to discriminate between the absence of cord and cord with stresses greater than certain maximum tolerable limits.

According to equation 4, tensile stresses generated by thermal shock depend on glass thickness. In this study, four glass thicknesses were considered (0.025", 0.050", 0.075" and 0.100") since they encompass the range of values for the sidewall of most non-refillable bottles. According to Figure 2, the thermal shock stress indexes (X in equation 5) are 13, 20, 25 and 29 for these glass thicknesses, respectively.

Considerations were also made in this study of three glass surface conditions associated with abrasions created from the use of 40 grit, 150 grit and 320 grit emery paper. The specific glass strength values, for a load duration of 2 seconds typical of thermal shock, were determined through previous studies and are 3000 psi, 4300 psi and 5200 psi, respectively. Finally, thermal shock test differentials of 90°F, 100°F and 110°F were used as these are convenient values obtainable by commercially available thermal shock test equipment.

Calculations were made of the minimum magnitude of tensile cord stress that will result in failure during an abraded thermal shock test using these parameters and equation 6. The results are summarized in Tables II, III and IV for various surface conditions and thermal shock test differentials of 90° F, 100° F and 110° F, respectively. As shown by these data, the minimum cord stresses that can be detected by these tests depend strongly on glass thicknesses. For example, for a temperature differential of 90° F (see Table II) and a glass surface condition associated with a 40 grit emery cloth abrasion, bottles that fail the test would indicate a cord stress ranging anywhere from 570 psi to 1830 psi, depending on glass thickness. Similar results are shown for other abrasions and for other thermal shock differentials as shown in Tables III and IV.

Thus, without taking glass thicknesses into account, the test would indicate the presence on the outside glass surface of cord stresses ranging from negligible levels (less than 500 psi) to levels of serious cord stress magnitudes (greater than 1500 psi). Such disparities make the test worthless and mis-leading. However, if the glass thickness at the fracture origin is measured and taken into account, then the results can be used to estimate the magnitude of the cord stress with some degree of reliability.

The data in Tables II, III and IV are graphically shown in Figures 3, 4 and 5 for various surface conditions and thermal shock test differentials of 90° F, 100° F and 110° F, respectively. These plots can be used to determine the utility of using various test conditions (abrasions and thermal shock differentials) to detect cord stresses of various magnitudes such as 500 psi, 1000 psi, 1500 psi and 2000 psi. For example, a tensile cord of 500 psi acting on the outside glass surface would result in breakage in a 90°F thermal shock test only for a glass thickness of 0.100 inches and an abrasion of 40 grit. Similarly, a tensile cord of 2000 psi would result in breakage in a 90°F thermal shock test for all glass thicknesses with an abrasion of 40 grit and for a glass thickness of greater than 0.075 inches and an abrasion of 150 grit. The test will not work for a 150 grit abrasion and glass thicknesses less than 0.075 inches and for all glass thicknesses and an abrasion of 320 grit.

For convenience, the interpretation of the thermal shock tests are summarized in Tables V, VI and VII for cord stress levels of 500 psi, 1000 psi, 1500 psi and 2000 psi using the plots in Figures 3-5. Plots and tables similar to these would have to be constructed in order to properly interpret the results of abraded thermal shock tests.

CONCLUSIONS

Based on the calculations shown in this paper, it was concluded that without taking glass thickness at the fracture origin into consideration, the results of abraded thermal shock tests to quantify the magnitude of tensile cord streaks that are on the surface of test containers are worthless and potentially misleading. In order to use this type of test, it is imperative that the glass thickness at the fracture origin be measured and taken into consideration.

REFERENCES

1. Standard Terminology of Glass and Glass Products, ASTM C162, West Conshohoken, PA.
2. Stones and Cord in Glass, C. Clark-Monks and J. M. Parker, Society of Glass Technology, Sheffield, England, 1980, page 130.
3. Standard Test Method for Photoelastic Determination of Residual Stress in a Transparent Glass Matrix using a Polarizing Microscope and Optical Retardation Compensation Procedures, ASTM C978, West Conshohoken, PA.
4. V. C. Zwicker, Ring Section Examination of Glass Containers, Bulletin Am. Cer. Soc., Vol. 18, No. 4, April, 1939, pp 143-147
5. Glass Container Association, Polariscopic Examination of Glass Container Sections, J. Am. Cer. Soc., Vol. 27, No. 3, March, 1944, pp 85-89

TABLE I: Some Examples of Cord Sources

Raw Materials
    Placed in wrong storage area
    Wrong composition
    Impurities
    Wrong particle size

Batching Operations
    Weighing errors
    Batch segregation
    Inadequate batch mixing

Furnace Issues
    Refractory Erosion
    Changes in flow patterns caused by firing patterns or pull rate
    Raw material dusting problems

TABLE II: Minimum Detectable Cord Stress (psi) for a 90° F Thermal Shock Differential

Glass Thickness (inches)	Stress Index	40 grit abrasion (3000 psi)	150 grit abrasion (4300 psi)	320 grit abrasion (5200 psi)
0.025	13	1830	3130	4030
0.050	20	1200	2500	3400
0.075	25	750	2050	2950
0.100	29	390	1690	2590

TABLE III: Minimum Detectable Cord Stress (psi) for a 100° F Thermal Shock Differential

Glass Thickness (inches)	Stress Index	40 grit abrasion (3000 psi)	150 grit abrasion (4300 psi)	320 grit abrasion (5200 psi)
0.025	13	1700	3000	3900
0.050	20	1000	2300	3200
0.075	25	500	1800	2700
0.100	29	100	1400	2300

TABLE IV: Minimum Detectable Cord Stress (psi) for a 110° F Thermal Shock Differential

Glass Thickness (inches)	Stress Index	40 grit abrasion (3000 psi)	150 grit abrasion (4300 psi)	320 grit abrasion (5200 psi)
0.025	13	1570	2870	3770
0.050	20	800	2100	3000
0.075	25	250	1550	2450
0.100	29	*	1110	2010

* container will fail from thermal shock stress alone

TABLE V: Cord Detection Capability using a 90° F Test Differential

40 grit abrasion (3000 psi)	Glass Thickness (inches)	Stress Index	Cord = 500 psi	Cord = 1000 psi	Cord = 1500 psi	Cord = 2000 psi
	0.025	13	N	N	N	Y
	0.050	20	N	N	Y	Y
	0.075	25	N	Y	Y	Y
	0.100	29	Y	Y	Y	Y
150 grit abrasion (4300 psi)	0.025	13	N	N	N	N
	0.050	20	N	N	N	N
	0.075	25	N	N	N	?
	0.100	29	N	N	N	Y
320 grit abrasion (5200 psi)	0.025	13	N	N	N	N
	0.050	20	N	N	N	N
	0.075	25	N	N	N	N
	0.100	29	N	N	N	N

TABLE VI : Cord Detection Capability using a 100° F Test Differential

40 grit abrasion (3000 psi)	Glass Thickness (inches)	Stress Index	Cord = 500 psi	Cord = 1000 psi	Cord = 1500 psi	Cord = 2000 psi
	0.025	13	N	N	N	Y
	0.050	20	N	Y	Y	Y
	0.075	25	Y	Y	Y	Y
	0.100	29	Y	Y	Y	Y
150 grit abrasion (4300 psi)	0.025	13	N	N	N	N
	0.050	20	N	N	N	N
	0.075	25	N	N	N	Y
	0.100	29	N	N	Y	Y
320 grit abrasion (5200 psi)	0.025	13	N	N	N	N
	0.050	20	N	N	N	N
	0.075	25	N	N	N	N
	0.100	29	N	N	N	N

TABLE VII: Cord Detection Capability using a 110° F Test Differential

40 grit abrasion (3000 psi)	Glass Thickness (inches)	Stress Index	Cord = 500 psi	Cord = 1000 psi	Cord = 1500 psi	Cord = 2000 psi
	0.025	13	N	N	N	Y
	0.050	20	N	Y	Y	Y
	0.075	25	Y	Y	Y	Y
	0.100	29	Y	Y	Y	Y
150 grit abrasion (4300 psi)	0.025	13	N	N	N	N
	0.050	20	N	N	N	N
	0.075	25	N	N	?	Y
	0.100	29	N	N	Y	Y
320 grit abrasion (5200 psi)	0.025	13	N	N	N	N
	0.050	20	N	N	N	N
	0.075	25	N	N	N	N
	0.100	29	N	N	N	?

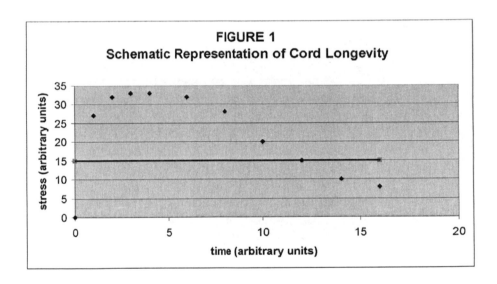

FIGURE 1
Schematic Representation of Cord Longevity

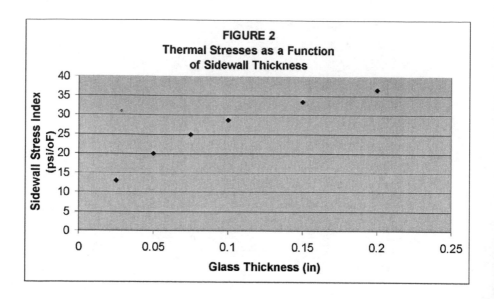

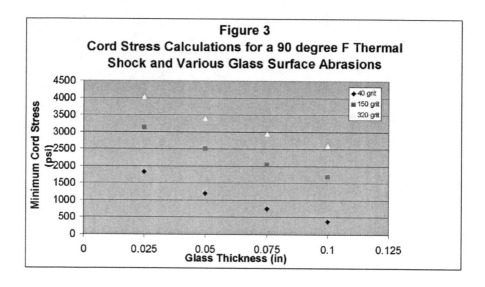

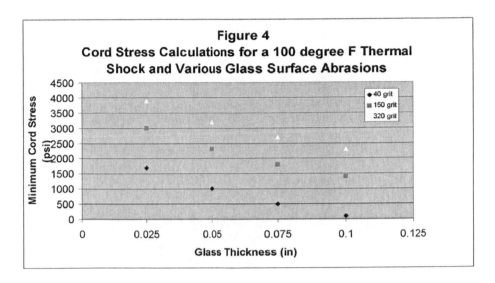

Figure 4
Cord Stress Calculations for a 100 degree F Thermal
Shock and Various Glass Surface Abrasions

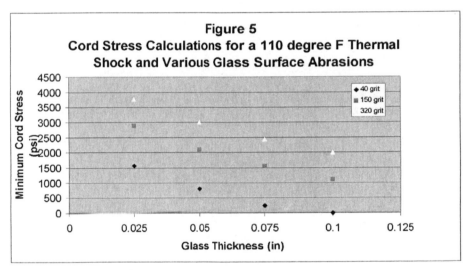

Figure 5
Cord Stress Calculations for a 110 degree F Thermal
Shock and Various Glass Surface Abrasions

TO WET OR NOT TO WET, THAT IS THE QUESTION – PART B – USING DRY BATCH

Douglas H. Davis and Christopher J. Hoyle
Toledo Engineering Co., Inc.

I.   ABSTRACT

Alternative methods of preventing batch segregation and carryover that do not carry the energy penalty of water-wetting are reviewed. Options for suitable charging of dry or agglomerated batch are also discussed.   Changes in charging are suggested to provide suitable log formation and batch recirculation with dry batch. Changes to promote early fritting should reduce the carryover from the dry batch.

Oil wetting is discussed as giving many of the benefits normally associated with water-wetting. Batch agglomeration without waste gas preheating is not seen as a suitable substitute for water-wetting, having a similar energy situation.

Preheated, agglomerated batch avoids the segregation problems of dry batch and energy penalty of water-wetting. The binding of particles and rapid melting minimize the carryover. Suggested changes to the charging system should alleviate problems with the flat, foam-covered batch blanket reported with agglomerated, preheated batch.

II.   INTRODUCTION

Wanting to avoid the energy cost of evaporating water used in batch wetting is logical. In Part A of this paper[1], arguments were made, however, that using dry batch without special adaptations was likely to be a financial mistake. Batch segregation, increased carryover, and charging problems were expected as a result of using dry batch, leading to added expenses from shortened furnace life and increased energy use to restore glass quality. These were predicted to overshadow energy savings from dry batch.   However, the need to develop techniques for use of dry batch was acknowledged as batch preheating comes into more general use.

In this part B, we review methods of preventing batch segregation and carryover other than water-wetting. Options for suitable charging of dry or agglomerated batch are also discussed.

III.   ISN'T THERE ANY WAY OTHER THAN WATER-WETTING?

Several liquids can be used as alternatives to water, i.e. caustic soda, and fuel oil, but do not generally provide a cost savings. TECO has long advocated batch agglomeration, in conjunction with preheating with waste heat. Agglomeration addresses the concerns we discussed in part A, in addition to facilitating batch preheating. There are also mechanical techniques of charging that may provide assistance.

These various alternatives will be discussed below, grouped to correspond to the glassmaking problems from dry batch addressed in Part A, i.e.

- Segregation,
- Carryover and dusting, and/or
- Providing a recirculating batch pattern.

IV.   ALTERNATE MEANS TO ACHIEVE BENEFITS OF WET BATCHING

A.   Preventing Segregation

1.   Alternative Wetting

a.   Caustic Soda Solution

One alternative to water-wetting is to use caustic soda (NaOH solutions) as a partial substitution for the alkali of soda ash; most commonly a 50% solution of NaOH in water. This is not helpful in terms of dry versus wet batch. Water is still being added and must be evaporated. In addition, caustic-wetting is more expensive overall than water-wetting, as 12-15% of the batch soda ash will be replaced with the generally higher priced alkali from caustic soda.

b.   Oil-Wetting of Batch

Some glassmakers have wetted mixed glass batch with oil in the past to avoid the "drying out" caused by soda ash hydration during batch storage. The oil does not interact chemically with batch ingredients and remains as a wetting liquid until it is burned off in the melter. Therefore it minimizes segregation, facilitates log formation at the charger, plus reducing batch dusting in the plant. The wetting oil will be combusted within the melter and while not burned as efficiently as the main melter fuel, all of the potential energy will be released inside the melter.

Oil in the batch will influence the refining system. Some residual organic solids (char) should be formed in the batch and retained to influence the sulfate refining. When beginning oil-wetting, excess oxygen values and glass redox should be monitored for any required adjustment. The odor of the oil will be a consideration. Heavy oil will be less volatile, but may require heated tanks.

Oil-wetting of the batch will not assist with minimizing carryover or dusting in the melter. In addition, if preheating is initiated, the oil will be burned off and its' benefits will cease.

A.   Preventing Segregation (Cont'd)
2.   Agglomeration of Batch

Presenting agglomeration as an alternative to water-wetting is somewhat misleading. Agglomeration accomplishes the goals of water-wetting, but does not eliminate the energy penalty from water evaporation. Agglomerating requires batch moisture and usually binders for cohesion. However, we are assuming that batch agglomeration is done as a package with a preheating system, so that the energy to drive off water and binders is waste energy from the furnace exhaust, not energy purchased from the outside. Thus batch agglomeration can reasonably be considered an alternative to water-wetted batch.

TECO has been an advocate of glass batch agglomeration in some form. Agglomeration will lock in the desired batch mix, preventing segregation. Both pelletization (via tumbling) and briquetting (by mechanical compaction) have been used in glass industry trials in the past.

Getting to batch preheating is critical to the glass industry and agglomerated batch will assist with direct preheating as the spaces between agglomerates will improve flow of the hot gases and convective heat transfer. The bonding will also resist incorporation of fine particles and dust into the waste heat stream, even after preheating.

With agglomeration, opportunities for raw materials open up significantly, allowing use of very fine sands, sands with major fractions below 200 mesh, very fine burnt limestone and dolomite, etc. Cost savings from easier melting, less expensive materials, shipping from closer deposits may well pay for the agglomeration. Of course, this added agglomeration step is another expense and, like preheating itself, has been difficult to justify as energy prices stay at lower than expected prices.

A.   Preventing Segregation (Cont'd)
3.   Mixing At Melter

In most glass plants, batch weighing and mixing is done at the raw material storage site, located at some distance from the furnace. The mixed batch is then transported via elevators, chutes, belts and perhaps even pneumatic transport to a bin feeding the furnace. Unless skillfully engineered, with a dry, flowable batch, each transfer point introduces some level of de-mixing. Glassmakers for years have depended on water-wetted batch to resist this segregation. The batch is kept moist long enough to retain a moist consistency at the furnace charger.

Without batch wetting or agglomeration, a workable alternative is to locate the batch mixer above the furnace daybin. The individual raw material weighments are collected at the storage area and then transported to the mixer. Even pneumatic transport works well in this situation, any segregation being eliminated at the mixer. Early opposition to this arrangement was due to the difficulty of mixer repairs at a several-story elevation. However, modern mixers are more reliable.

A   Preventing Segregation (Cont'd)
4.   Mass Flow Bins / Chutes
       With wet batch minimizing segregation effects, and a drive to cut batch facility cost, designing for segregation avoidance has not a big priority. If use of dry batch is to be considered, all bins and chutes need to be carefully reviewed for mass flow consideration. In addition, various re-blending devices may be needed.

A.   Preventing Segregation (Cont'd)
5.   Matching Particle Size of Raw Materials
       Segregation can be minimized by matching granulometry of the different raw materials. Particle segregation in flowing material is a function of both particle size and density. We have little control over density, but minimizing the differences in particle sizes in the mixed batch is possible. This was practiced by some glassmakers before the industry turned to wet batch. If we return to dry batch, this could be important again. This can be expensive; at times requiring extra transport costs and premium charges for special processing. It also can run counter to desired sizing for decrepitation, and melting rates.

IV.   ALTERNATE MEANS TO ACHIEVE BENEFITS OF WET BATCHING (CONT'D)
B.   Minimizing Dusting / Carryover in Melter
       In Part A, the bonding of fine particles by dried soda ash solution was discussed, minimizing the pickup of batch particles by the fires and deposition on furnace walls, crown, and the regenerators. Without water, other tools need to be considered.

1.   Changing Raw Materials
       Even with wet batch, raw materials should be picked with concern for carryover. As stated before, TECO has determined that sands with more than 7% fines below 140-mesh are connected with furnace life problems (even with wetted batch). Similarly, alkaline earths (dolomite, limestone) should exhibit decrepitation values of less than 5%. If dry batch is to be used, it may be wise to change raw materials (if at minimal cost). Most important will be the choice of sand and dolomite.
       For dry batch, the sand should be as coarse as the melter size and tonnage will allow (perhaps 25-30 mesh) and with minimum fines below 100 mesh. Such sand will reduce carryover, but may limit the tonnage that can be pulled at high quality levels.
       Unless the supply of limestone/ dolomite shows very low decrepitation (less than 1%), the alkaline earth chosen should be very coarse (6-10 mesh max), with no material below 60-80 mesh. This will reduce the decrepitation effect. Decrepitation is enough of a problem that glassmakers commonly pay extra transport for low-decrepitating dolomite.
       Moving to an agglomerated batch opens raw material options both in terms of cost and improving melt rate. Very fine sands, fine burnt limestone and dolomite, etc. not only offer melting advantages but improve the strength of the agglomerate.

B.   Minimizing Dusting / Carryover in Melter (Cont'd)
2.   Agglomeration Of Batch
       Bonds developed during agglomeration will also resist the pickup of individual particles by combustion gases. Batch agglomeration is only considered here in combination with preheating with

waste exhaust gases. Without this re-use of waste energy, batch agglomeration (requiring water and additives) is still at an energy disadvantage compared to dry batch.

The preferred situation is that batch would be agglomerated, preheated, and charged directly from the preheater into the melter at a temperature of 600-700 C. Entering a furnace atmosphere of 1400-1600C, the surface of these agglomerates should begin to react and fuse very rapidly (fritting). This will reduce the particulate pickup even more.

B.   Minimizing Dusting / Carryover in Melter (Cont'd)
3.   Rapid Fritting at Melter Entrance

Where high levels of carryover are expected (certainly with dry batch), the charging area design should be carefully considered. The goal would be early melting of the batch surface to minimize pickup and carryover by the flames in the melter. The term "fritting" is used, although somewhat of a misnomer.

Figure I a
Float Furnace –
Rear Wall to 1st Port

Technical compromises are required in this area. The batch surface needs to be fused rapidly, but high velocity flames blasting over newly charged batch can aggravate instead of minimize carryover. It is important not to overheat the underlying glass in the rear of the tank, because significant thermal gradients from the hot spot to both ends of the furnace are needed to develop the critical convective glass flows in the bulk glass. However, in the doghouse areas, the atmosphere above the batch is quite isolated thermally from the glass below the batch by the batch and foam. We also need to recognize that fritting will restrict losses of gases from the batch; impeding refining somewhat, but modeling shows that refining happens faster than melting. For discussion purposes, the following suggestions for faster "fritting" will be visualized on a side-port regenerative furnace.

3.   Rapid Fritting at Melter Entrance (Cont'd)
a.   Manipulation of the Waste Gas Recirculation Patterns

Circulation of the combustion gases inside the melter superstructure can transfer the heat into this "fritting" zone somewhat gently. Numeric modeling predicts and displays the flow of combustion gases from the firing ports. The bulk of the gases travel directly from firing port to exhaust port, but a significant volume does not enter the exhaust port, circulating instead around the furnace. Both figures 1a and 1b shown here are sections of float furnace models, from the rear suspended wall to just past the first port.

This effect has been used as an adjustable design tool for distributing heat. The port and burner design has only a minimal effect on this recirculation, but the space made available for the recirculation does have a significant effect.   A full-width and deeper (longer) doghouse will strengthen the recirculation and increase the transfer of heat to the doghouse and the batch.

The velocity of these circulating gases is quite low, and should not aggravate carryover. Increasing doghouse depth (length) must be done carefully, taking care that the rearward-flowing hot glass stream below the batch does not descend early and not carry adequate heat to the bulk glass below the doghouse.

3. Rapid Fritting at Melter Entrance (Cont'd)
b. Use of Rear Wall / Superstructure as a Radiative Shell

The superstructure shape in the doghouse area should be designed to improve radiative heating of the batch rather than relying mainly on convective heating from waste gases or flames moving over the batch. The cold batch provides an ideal heat sink for the radiant energy emanating from the sloping part of the back wall, the crown, and breastwalls. Radiant energy transfer is, of course, driven by the fourth power of the temperature gradient as $Q = \sigma A \epsilon (T_2^4 - T_1^4)$. This is an area worthy of investigation to see if improvements can be made. Maximum use must be made of the available radiant energy.

The numerical model of a float glass tank was adapted to evaluate the effect on heat distribution to the batch of several possible adjustments to the doghouse area. These included application of a high emissivity coating to the rear interior wall and adjustment of the suspended wall geometry to reflect more energy to the batch. Increased height and volume of the shell over doghouses to increase batch heating has been discussed recently in workshops and in the literature [2, 3]... While we cannot take credit, it is a logical move.

3. Rapid Fritting at Melter Entrance (Cont'd)
c. Use of Flat-Flame Radiant Burners on Suspended Wall

The suspended wall filling in the doghouse area of a side-port furnace can be used as a support for a bank of flat-flame radiant burners. These can be mounted within several feet from the cold batch surface, greatly supplementing the radiant heating of the incoming cold batch. These burners will provide almost no added gas velocity at the batch surface.

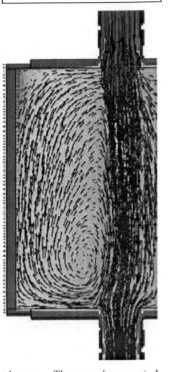

Figure I b
Float Furnace –
Rear Wall to 1st Port

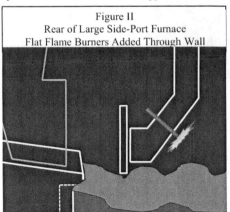

Figure II
Rear of Large Side-Port Furnace
Flat Flame Burners Added Through Wall

It would also be possible to provide a horizontal shelf on these suspended walls to bring all the burner flames the same distance from the batch surface.

3. Rapid Fritting at Melter Entrance (Cont'd)
d. Oxy-Fuel CGM (Convective Glass Melting)

In 1995, BOC introduced their vertically-mounted oxy-fuel burners (CGM). Initial applications included various fiberglass tanks and on-the-fly boosting of impaired furnaces. These were designed to provide intense convective heating of the batch surface, emit low levels of NO$_x$ and provide flexible application. In early test work, very low levels of batch entrainment were noted. Applications for batch "fritting" were expected. However, information from some applications indicated refractory damage and carryover into regenerators. Producers of boron-containing glasses have anticipated unacceptable volatilization of boron.

3.   Rapid Fritting at Melter Entrance (Cont'd)
          e.   Port Zero Oxy-Fuel Burners

Figure III
Oxy-Fuel Port Zero Burner
Added to Float Tank

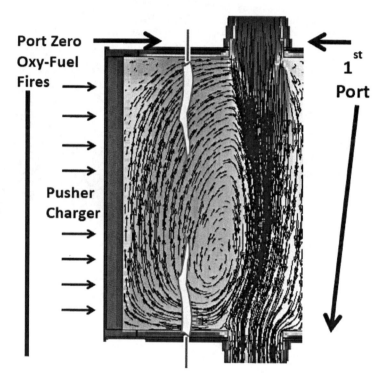

A growing application on side-port melters is addition of supplementary oxy-fuel burners between the first firing port and the back wall (charger). Useful for helping with partially plugged checkers or just for a tonnage boost, these are commonly called "port zero" burners. The flame velocities of these fires are quite low, due both to burner design and the lower combustion volumes from oxy-fuel burners. These can be directed to help heat the batch surface.

f. Separate Fritting Chamber

A conclusion from the above discussion might be to provide a closed chamber at the charging end of the furnace with burners designed just to "frit" the batch surface. A suspended curtain wall could separate this chamber from the melter proper.

There are several reasons why this would be a last choice. Requiring additional floor space, this would probably have to be a green-field furnace, not being practical for a rebuild nor an on-the-fly modification. Also, suspended curtain walls are questionable for campaigns 15 years or longer. A dropped section of curtain wall is a difficult fix.

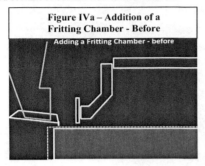

Figure IVa – Addition of a Fritting Chamber - Before

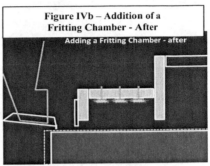

Figure IVb – Addition of a Fritting Chamber - After

IV.   ALTERNATE MEANS TO ACHIEVE BENEFITS OF WET BATCHING

C.   Recirculating Batch Patterns

Charging systems are of three general types. The first, on relatively small furnaces, consists of one or more screw chargers. Another system, used for somewhat larger furnaces is the small sidewall pusher chargers, or rabble chargers, commonly of the GANA-type. The third is the very wide end-wall pusher "blanket" chargers, capable of feeding large tonnages. Many of this type are supplied by Merkle.

As discussed earlier in Part A, the charging pattern using screw chargers is fairly independent of this wet vs. dry batch discussion. However, the pattern obtained on the glass surface using either the sidewall/rabble chargers or the end wall pusher, is very dependent on the batch consistency, i.e. wet vs. dry. Consistent with the premise that dry batch (preheated) will be necessary; it is important to determine how to create the batch patterns the industry has learned to love from wet batch, but now with dry batch.

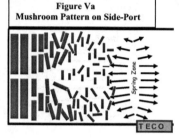

Figure Va
Mushroom Pattern on Side-Port

As shown in Figures Va and Vb, the mushroom or horseshoe pattern on large side port furnaces (end wall, pusher, or Merkle-type chargers) is quite different from the wheel batch patterns on successful oxy-fuel and end-port furnaces (sidewall, rabble, GANA-type chargers).

Figure Vb
Wheel Pattern on End-Port

However, the critical feature of their circulation is the same, with residual piles on the end of the pattern moving back toward the melter rear wall under the influence of the rearward glass flow. In the rear of the furnace, melter residuals are incorporated with fresh batch, and given a second chance at melting and homogenization.

This batch circulation is critical to reaching the optimum equilibrium between quality, tonnage, and energy usage. Most medium to large tonnage melters (wide pusher chargers) are currently charged using water-wetted batch, and moist consistency is the key to log formation. The no-flow properties of wet, cohesive batch leads to the retention of splits between charged piles. These splits open up to give free-floating batch piles, which can respond to the convective flows in the bath, and give us the desired recirculation.

If the batch is changed from a wet to dry, the dry batch will flow (low angle of repose) to fill in the breaks between pushes and a flatter batch blanket is formed. This flat blanket acts like a single mass, tapering to a thin foamy edge. Without small logs breaking off, the melting residuals are retained on the front edge and will either melt out or pass into the product stream. Unless tonnage is very low, glass quality will be hurt. In addition, the flatter batch blanket provides no low areas for foam to flow into, in order that cold batch is exposed to the flames. The reflective foam reduces heat transfer from the flames. This reinforces the expectation of reduced glass quality from dry batch.

Log formation needs to be restored – but with dry batch.

The term "dry batch" has been used casually. Three things could be meant by "dry batch". The first would be normal mixed batch with no water-wetting. As discussed in Part A, TECO considers normal mixed dry batch (flowable) to be a bad financial choice for larger regenerative furnaces.

The second "dry batch" is the preferred combination, i.e. batch that is both agglomerated and preheated. Here, improved charging will be very important. With preheated, agglomerated batch, the hot agglomerates immediately slump, softening as they enter the furnace. The resulting blanket will be a reasonably flat, featureless blanket with little tendency to form logs. The blanket would be much shorter, however, with more time for homogenization. Predicting the overall effect on quality is difficult. However, it is safe to say that forcing log formation by altering the charging mechanism would be a benefit.

The third type of "dry batch" would be batch that is agglomerated but not preheated. This intermediate situation will have an angle of repose between wet and dry and the break between logs should be indistinct. Improvements in the charging will be important here.

Thus a change to dry or agglomerated batch (pre-heated or not) calls for some change in charging techniques to create batch logs with distinct separations. There seem to several viable approaches.

IV.   ALTERNATE MEANS TO ACHIEVE BENEFITS OF WET BATCHING (CONT'D)
C.   Recirculating Batch Pattern (Cont'd)
1.   Variation on Standard Pusher Chargers
Dry batch creates the most severe problem on the large side-port regenerative furnaces, where the traditional charger is the rear-wall blanket charger. These chargers feed a blanket of batch over the furnace width. Movement of the blanket by the charger and

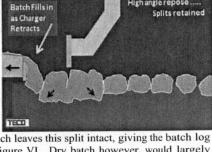

Figure VI -Formation of Logs with Pusher Charger and Wet Batch

Figure VII - Alteration of Charging to Leave Gap and Create Glass Spacer (a-e)

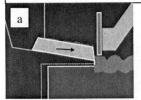

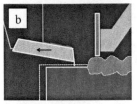

subsequent   release causes   a   rise,   slight rotation,   and   formation of a split in the blanket. The   steep   angle   of repose of water-wetted batch leaves this split intact, giving the batch log boundaries as shown on figure VI. Dry batch however, would largely fill these splits in, minimizing log definition.

A suggested change to generate log formation with dry or agglomerated batch would be to 1) use a significantly longer "push" into the furnace (12-18") and 2) to delay dropping new batch onto the glass surface until the charger has partially retracted. This sequence is illustrated in Figures VIIa through VIIe. This would create an initial hot gap (uncovered glass) between the last batch pile pushed and the batch pile waiting to be pushed. Any batch pile on the molten glass, of course, sinks partially into the molten glass, analogous to an iceberg. Therefore, with the next charger push, the two piles cannot be

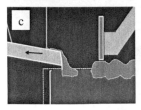

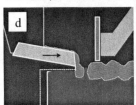

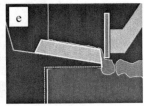

completely pushed together; a small molten glass "wave" creating a spacer.

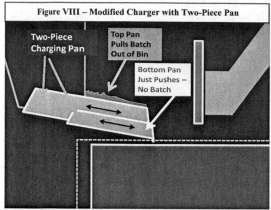

Figure VIII – Modified Charger with Two-Piece Pan

Two-Piece Charging Pan

Top Pan Pulls Batch Out of Bin

Bottom Pan Just Pushes – No Batch

Delayed charging of batch onto the surface, leaving an uncovered glass gap, requires modification of the charger pan. The normal single-piece charger pan would need to be replaced with a two-piece charger pan as shown on Figure VIII. The second portion of the charger pan would be located beneath the upper normal pan, and would not extend until the upper pan is fully extended. The lower pan section (with a pusher "nose") would then extend 6-12" further, pushing the batch pile further into the furnace. This lower pan section would have no batch on it, of course, and in retracting to its storage position beneath the upper pan, would drop no new batch on the glass. But as the upper charging pan then retracts to its' rearward position, new batch would be dropped on the glass as normal. However, an open uncovered gap on the glass surface would be left.

Several potential problems need to be considered. One is that the charger has to deal with higher temperatures. In addition, the free-flowing dry batch will make lower batch piles than wet batch so that the same charged tonnage with dry batch (although formed into logs) would logically extend farther down the tank; possibly beyond the hot spot. However, the thinner dry batch piles should melt out and disappear faster, shortening the blanket and compensating. If the "dry batch" is agglomerated, past work has shown that the piles will melt faster, shortening the blanket compared to a non-agglomerated batch. In addition, if the agglomerated batch is pre-heated, the blanket will be shortened even further. A longer blanket is unlikely.

A more serious concern is that the dry batch blanket, although broken into discrete logs, will be quite flat with foam covering most of the area, making heat transfer from the flames difficult. This will be especially true with the rapidly melting preheated batch. One possibility is the widely reported use of reducing atmospheres at the glass surface to break down the foam. It would be practical to structure the gas/air introduction so that a reducing atmosphere is present over much of the bottom of the fires.

C. Recirculating Batch Pattern (Cont'd)

2. Alternate Charging of Cullet and Batch

If a volume of cullet is charged between volumes of batch, this cullet will take time to soften, and will maintain spacing between batch logs. The cullet spacers will melt out much faster than the batch, however, releasing the individual logs to respond to the convective flows. This is an extension of the discussion immediately above, where a two-stage charging pan would leave an uncovered hot glass space gap, and then a small hump of soft glass would be pushed between batch piles. However, in this case the open space on the glass surface would be filled from above with crushed cullet. A secondary bin containing only cullet would discharge a 3-4" layer of cullet into this uncovered space. This is shown on the three sections of Figure IX.

A concern that has been expressed is that in float glass particularly, the separated charging of cullet as "spacers" may consume all of the available cullet. In float plants, 15-20% cullet levels are often the upper limit since outside cullet is used only with great care. The anticipated problem here is

Figure IX – Separate Feeder for Cullet Spacer

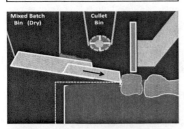

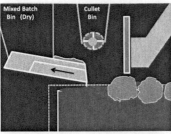

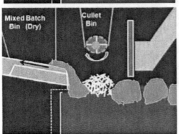

that the cullet being separated from the batch may lead to glass inhomogeneities. One counter argument is that Univerbel feeders laid cullet as a separate layer below the raw batch and apparently did not cause a homogeneity problem. Recent review articles [4, 5] make the case that cullet in glass actually retards the rate-determining dissolution of sand grains. Therefore if we separate the cullet from the batch (using cullet "spacers"), it appears that faster sand dissolution should override reduced heat penetration into the batch and the possibility of cullet sinking. One mention of problems with cullet was that variation of cullet levels led to major temperature variations in the furnace. This does not apply here since the cullet level will remain steady.

References to poorer refining with high cullet levels might imply that pockets of pure cullet might give areas of poorly refined glass. This should be countered by having all or most of the cullet on the melter surface.

C.   Recirculating Batch Pattern (Cont'd)

3.   Screw Chargers with Pushers

Screw chargers have shown themselves above the fray with respect to dry or wet batch. The charging pattern is simply a straight line stream, until some interference (tank wall, other batch, glass current) causes the end of the string to deflect.

With this random behavior, screw chargers are not very good at creating a consistent pattern. A hybrid between a screw and a charger is a suggested solution.. This not original to us, but is a good idea [6].

For medium to low tonnage furnaces, TECO suggests batch be fed with multiple screw chargers, each fitted with a periodic pusher. This pusher would periodically drop, cut into the batch stream close to the wall, push the forward batch section 12-18" downtank, rise up and return to the rest position above the screw charger. This pusher action should create a temporary open gap on the glass surface that would not be completely closed up at the next "push". The residual gap would be partially due to the hot glass carried ahead by each "pushed" pile forming a small pad between batch piles. This is shown in Figures Xa-Xe.

This combination of screw plus pusher has the traditional advantage of screws being sealed from dusting, plus producing logs without wet batch. Multiple feeders with adjustable feed rates and adjustable "push" lengths will allow coordination with the convective glass flows to create the desired recirculation pattern on the glass.

Figure X
Screw Charger Plus Pusher

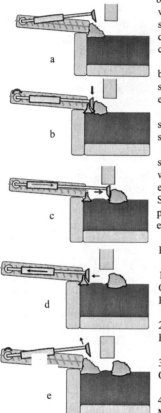

a

b

c

d

e

IV.   ALTERNATE MEANS TO ACHIEVE BENEFITS OF WET BATCHING (CONT'D)

C.   Summary – Reasonable Alternatives to Wet Batch

Oil wetting should be a practical way to obtain many benefits associated with water-wetting. The heat value of the oil will be released inside the melter, so that much of the energy savings from avoiding water use could be realized. Carryover and dusting will not be controlled with oil, so that the early fritting changes should be implemented.

Using dry, flowable batch, suggested changes to the charging systems should provide suitable log formation and batch recirculation. Changes to promote early fritting should reduce the carryover from the dry batch.

Batch agglomeration without waste gas preheating is not a suitable substitute for water-wetting, having a similar energy situation.

Use of preheated, agglomerated batch would avoid both the segregation problems of dry batch and higher energy costs of water-wetted batch. The binding of particles and the rapid melting upon entering the furnace should minimize carryover problems. Suggested changes to the charging system should help alleviate problems with the flat, foam-covered batch blanket normally expected with agglomerated, preheated batch.

References:

1.   DH Davis, CJ Hoyle, "To Wet or Not to Wet – That is the Question – Part A, Proceedings of the 71st Conference on Glass Problems, October 19th and 20th, 2010.

2.   R. Sims, GLASS WORLDWIDE, "Batch Preheating Breakthrough", V. 30, December/January 2011, p.32

3.   M. Lindig, GMIC Workshop on Waste Heat Utilization, October 2011, Columbus, Ohio

4.   Gelnar, S. and Smrcova, E., Glass International, v. 33, No.2, March, 2010

5.   Novotny, V., Glass International, v. 33, No,2, March, 2010

6.   M. Lindig, "Mathematical Modeling Study on a Charging Area Improvement of Glass Melting Furnaces", Proceedings of the 11th International Seminar on Furnace Design – Operation and Process Simulation, June 21-23, 2011, Velke Karlovice, Czech Republic

BATCH WETTING - ANOTHER POINT OF VIEW

John Brown[1], Hisashi Kobayashi[2], Wladimir Sarmiento-Darkin[2], and Matthias Lindig[3]

[1]Corning Inc, NY-USA,

[2]PRAXAIR Inc. Danbury-Tonawanda NY-USA,

[3]SORG, Lohr am Main, Germany

This paper will deal with water addition in batch from three different angles:

1. Use of water in batch and its impact on glass manufacturing cost
2. Fuel Reduction by Dry Batch and Dust Control by Low Momentum Oxy-Fuel Firing
3. Batch Wetting- Handling and Preheating

Each one of the authors wanted to reflect their own point of view on the subject and to analyze the critical parameters involved in the problem from their own perspective.

Part 1: WHY WE USE WATER IN BATCH AND HOW MUCH IT COSTS US
John Brown, Corning Inc, USA.

Improving energy efficiency in glass manufacturing process is of paramount importance for the glass industry. Energy intensity is closely followed by regulating agencies and consumers putting pressure in the industry to optimize and reduce consumption whenever and wherever is possible. Preheating batch and cullet seems like one of the technologies most promising to reduce energy consumption. However, water in the batch is a big hurdle that needs to be overcome before these technologies are widely accepted and implemented. Part 2 and 3 of this paper will touch in more detail all aspects related with batch and cullet preheating. My contribution will be to put a more global context to the opportunities lost by the introduction of batch wetting as a solution to batch dusting and carry over. Batch wetting was a temporary solution to a furnace batch dusting problem--that became permanent. Every gallon of water added to batch costs money. It is like putting dollar bills in the batch to prevent damage to the furnace. There are better ways and they are available today. This batched water prevents use of the largest potential energy saving approach.   That is incorporating waste stack heat in the preheating of batch and cullet.

Bill Manring, FMC corporation, published his approach to batching 4% water to control dusting[1]. Following the Oil Embargo of 1974, this was the beginning of the efforts of the DOE to make a major effort at reducing energy in every high energy intensive industry. Corning was selected in 1978 to be the bench mark for the US glass industry in energy reduction and I had the responsibility for Corning's 60 major regenerative furnaces. Adding water to batch was not part of my plan to reduce energy. After having the plant managers bonuses tied to their furnace energy performance, it became an impossible sell, from the melting departments to convince their bosses to add water to furnace fill.

Now, nearly 35 years later I find while doing furnace energy balances, that the use of water in batch is wide spread in the soda-lime glass melting segments. This is not a small segment as 68% of the tonnage of US glass is soda lime based container and about 14% float. A high proportion of this 82% of the total tonnage of glass is batching water and is worthy of concern. This represents a tremendous opportunity in cost reductions. The good news is that few are at the 4% water suggested, even when they think they are. I've measured from 1.75% to 5.25% but the majority in the 2 to 2.5%. This water requires heat to vaporize and additional sensible heat to reach the exhaust temperatures of the furnace. This is split about 40% for latent heat and 60% for sensible heat. Latent heat is non-returnable and a portion of the sensible heat is returned as preheated air. In a single reiteration for a 4% batched water example, the sensible heat amounts to about 0.75% as returnable heat to regenerators. Or, a net cost of batching this water of 5.25% of the purchased fuel. But, this is the tip of the iceberg. As said before, it is the batched water that prevents capturing the stack waste heat and returning as preheated batch and cullet.

Today, if I use $6 per decatherm of natural gas and 80% of the energy purchased by a melting plant is directed to primary melting, we are looking at an opportunity of $161 million, just for that portion of the US Glass industry that melts soda-lime glass. This is eliminating water in the batch and incorporating a batch and cullet preheating system that captures 33% of the waste stack energy and returns this heat to our cullet and batch feed stock. For this analysis I used the DOE's published figures of a quarter of a Quad of energy for the industry and used 82% which is the portion of the glass industry that is soda-lime and the prime user of water in batch. Additionally, I'm assuming 14% of this soda lime segment is oxy-fuel which identifies 23 billion CF of Natural Gas. The balance is regenerative and that amounts to 141 billion CF of NG. With a targeted savings for oxy-fuel of 25% and 15% for regenerative furnaces I arrive at an estimate, with $6/KCF of natural gas, of $161 million that is lost due to batching water and not incorporating waste heat in pre-heating batch and cullet.

The quarrel is not with Bill Manring, and his 1977 offer of water additions to prevent dusting, damage to breast walls and plugging checkers. Nor, is it with Doug Davis of TECO, who has shown the value of maintaining batch integrity with water to prevent plugging of regenerators and damage to breast walls. It is with our industry for not finding permanent solutions that don't add unnecessary costs of energy. This is a problem needing a solution that should have been identified in 1977, and solved in mid 80's. Getting the water out of batch is the first step in solving the problem of batch and cullet preheating. Several suppliers have cullet and some batch pre-heaters, and the jury is still out on some of these systems. Europe has little in country produced energy and has been more aggressive at promoting cullet and some batch preheating. Energy is just something that everyone needs to be thinking about and finding ways to be better users of energy.

A microscopic view of individual grains of batch grains and cullet provides an insight into a mechanism that few have seen. In 1968 Corning purchased a hotstage microscope from Bill Manring of FMC in Princeton. Yes, the same Bill Manring that offered batch wetting. Using 10 mil platinum foil and cutting and shaping to be a resistance heater, small grains of batch could be placed on the surface and films of the melting made along with the temperature sensed with a thermocouple tacked welded to the hot spot.

Unfortunately, I don't have these 8 mm films, those were made in 1968 and they are no longer available. However, Dr. Lindig had an example and we will show it in the figure below.

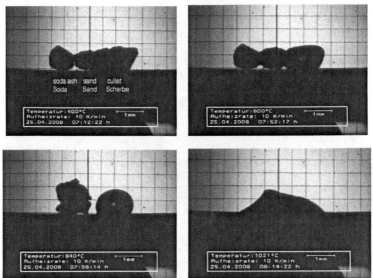

Figure 1. Soda ash, sand grain and cullet melting in a microscope. Courtesy of Prof. Dr. R. Conradt, RWTH Aachen. (1) Top left: T=400 C, t=to; (2) top right: T=800 C, t~40 min; (3) bottom left: T=840 C, t~44 min; (4) bottom right: T=1021 C, t~62 min.

The images above are very similar to those that convinced me back 1968 that water was not needed—if we used the heat in the furnace to pre-react the batch and protected the batch from the turbulent products of combustion. This example is a 28 second movie of a grain of soda ash, a grain of silica and a cullet chip. In figure 1 we have, (1) the grains at 400°C, a possible preheat temperature, (2) at 800°C when there is clear melting of soda ash, (3) at 840°C where the soda ash has covered the silica grain and is aggressively melting and the cullet has melted to a sphere, (4) at 1021°C when the melting of the three grains appears complete.

Part 2: FUEL REDUCTION BY DRY BATCH AND DUST CONTROL BY LOW MOMENTUM
       OXY-FUEL FIRING
Hisashi Kobayashi and Wladimir Sarmiento-Darkin, Praxair Inc, USA.

Batch wetting has been well established as the preferred method in the soda-lime glass industries to reduce dust during batch handling and carry over in the furnace, to improved batch homogeneity, to optimize batch pile geometry and to provide better alkali dispersion. Lehman and Manring[1] recommended to add about 4% H2O and to keep the batch at around 40 °C to maintain soda ash as monohydrate (Na2CO3.H2O). By keeping the temperature above 35.4 °C the amount of water lost on soda ash hydration is minimized. The effect of water addition was quantified by the angle of repose improvement obtained by wetting batch. Their experiment demonstrated a large increase in the angle of repose when the batch temperature was increased from 20 °C to 40 °C.

Lehman and Manring also analyzed dried samples of +20 mesh (0.84 mm) agglomerates formed in the batch after wetting. The samples were collected by screening the dried batch through a 20 mesh using a mechanical vibrator for 5 seconds. They found that the agglomerated particles were being bound together by recrystallized sodium carbonate compound. From their observations they concluded that significant amounts of soda ash dissolve in the wetting water and upon recrystallization this material provides the bounding medium. They speculated that the greater degree of soda ash dispersion plus the increased reactivity of the fine particles would increase the melting rate. However, this claim was not validated in the referenced work. Bonding of batch particles may increase heat conduction in batch piles and facilitate faster heating. Melting behavior of batch particles in the heated wire mesh screen shown in Figure 1 indicates rapid coating of sand particles by liquid soda ash once soda ash particles are melted at above 800 °C. Thus, dispersion of soda ash by batch wetting may not significantly enhance the melting rate.

While batch wetting has become a standard practice, it is highly desirable to come up with more efficient and cost effective methods to deal with these problems in view of the mounting social pressure to reduce CO2 emissions. Dusting and segregation during batch handling and transport could be controlled by improving the design of the existing batch handling systems. For dusting and carryover control inside the furnace oxy-fuel firing offers an attractive solution as the flame momentum can be reduced substantially by the elimination of nitrogen in the oxidizer stream and by improved burner designs using very low fuel and oxygen jet velocities. New oxy-fuel burners offered by Praxair are designed to operate with just a fraction of the momentum of the conventional oxy-fuel burners.

In order to determine the cost impact of adding water to the batch two independent modeling studies were performed for oxy-fuel float furnaces. Flue gases from most oxy-fuel fired furnaces exit at very high temperature without heat recovery, which makes the energy penalty of batch water addition for oxy-fuel fired furnaces higher than that for regenerative furnaces.

In the first study a detailed heat and mass balance of a 600 tpd oxy-natural gas fired float furnace with 15% cullet (based on glass produced) was completed. A parametric evaluation for different water contents in the batch was carried out at flue gas exit temperatures between 2400 °F and 2700 °F. Water content was varied between 0 and 4.5% of wet batch weight. Calculated fuel consumption is summarized in the following figure.

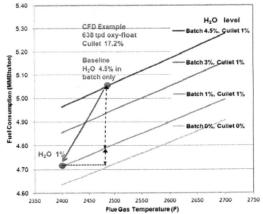

Figure 2. Oxy-fuel fired float furnace fuel consumption for different batch water additions.

For a constant flue gases outlet temperature 1% increase in batch water increases fuel consumption by about 1.6%. The vertical dotted line with arrows shows fuel consumption would decrease by about 5.6% when batch water level is reduced from 4.5% to 1%.

The second study was a three-dimensional computational fluid dynamic (CFD) modeling performed for a 638 tpd oxy-oil fired float glass furnace with 17.2 % cullet using the GFM software by Glass Services. Two batch water levels were evaluated, one with 4.5% and the second using only 1%. The firing rates of two burners near the batch charger were adjusted to match the glass exit temperature at the canal. The CFD study revealed that for the similar glass temperature profile, flue gas outlet temperature decreases when batch water level is reduced. In Figures 3a and 3b furnace gas temperature profiles are shown. The predicted flue gas temperature reduction was about 70 °F and the fuel savings was 7.9%. Using this CFD result the impact of batch water reduction with flue gas temperature reduction is shown in Figure 2 in the solid line with arrow connecting two points. The lower dotted line with arrows represents the additional fuel savings (1.5%) contributed from the flue gas temperature drop.

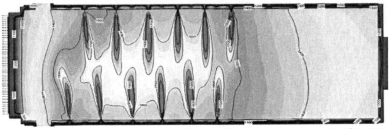

Figure 3a. Base Case, 4.5% H2O. Average flue gas outlet T= 1357°C (2475°F)

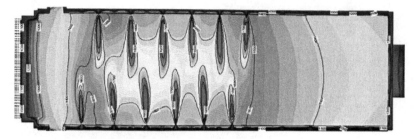

Figure 3b. Dry Batch, 1% H2O. Average flue gas outlet T= 1317°C (2404°F)

Lower flue gas outlet temperature by reducing batch water level may seem counter intuitive at first glance. Water present in the batch would act as a strong heat sink to reduce the temperature of flue gases before exiting the furnace. Although the cooling effect of the batch water may help to decrease flue gas exit temperature, the reduction of the firing rate near the flue ports and the reduced flue gas volume actually over compensates this effect, resulting in a lower flue gas exit temperature when batch water level is reduced.

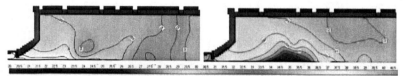

Figure 4. Water vapor volume fraction (%): Base case 4.5 % water (Left), 1% water case (Right)

Figure 4 shows a comparison of water vapor concentration near the batch feeder area. The high water concentration near the charger in the left figure indicates rapid evaporation of batch water. The very low water vapor concentration zone in the right figure is caused by CO2 evolution from soda ash and limestone.

The modeling performed above did not include possible faster melting rate of the batch particles bonded by recrystalized soda ash particles observed in the Lehman and Manring study. To evaluate this effect a third simulation was done with the CFD program with batch thermal conductivity doubled in the lower temperature range while maintaining the same conductivity at 1300 °C. (Note: thermal conductivity of batch materials increases sharply as soda ash starts to melt and react with sand at around 800 °C.) Results obtained with the higher thermal conductivity showed only about 14 °F drop in the flue gas exit temperature and little improvement on the fuel consumption. Thus, even a large increase in the effective conductivity of batch material by batch wetting does not appear to contribute much in reducing fuel consumption.

Fuel savings of 7-8% shown so far considered only batch water reduction and did not include the efficiency improvements by batch/cullet preheating using waste heat recovery. In Figure 5 flue gas flow rates by source and specific fuel consumption of a 600 tpd oxy-fuel fired float glass furnace with 15% cullet are compared for four scenarios including batch/cullet preheating to 932 °F (500 °C). The flue gas temperature was assumed to be constant at 2550 °F.

When the water in batch and cullet is removed completely, the flue gas flow rate decreases by about 14% (9%contributed by water removal and 5% by fuel input reduction) and the specific fuel consumption is reduced by 6.8%. If the air infiltration into the furnace could be eliminated, the total flue gas flow rate is reduced an additional 9% and fuel consumption comes down an additional 2.4%. Flue gas volume reduction creates an additional benefit downstream by reducing the size of air pollution control devices or waste heat recovery equipment. With batch/cullet preheating to 932°F, fuel consumption is reduced by 21.7% in this example. It is interesting to note that about 1/3 of the fuel reduction by batch/cullet preheating is due to batch/cullet water elimination.

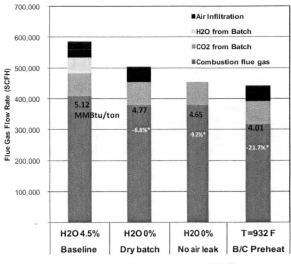

Figure 5. Flue gas volume by source and reduction of specific fuel consumption

One of the main reasons for batch wetting is to minimize dry batch carry over and plugging of regenerator checkers. Extensive studies[2] have been done to control build up of deposits in checkers by sulfates condensation, decrepitation of dolomite and limestone, and by batch dust carryover for air-fired regenerative furnaces. Although oxy-fuel fired furnaces do not suffer from the regenerator problems, dusting in the furnace is still a big concern for refractory corrosion as well as for increased particulates emissions.

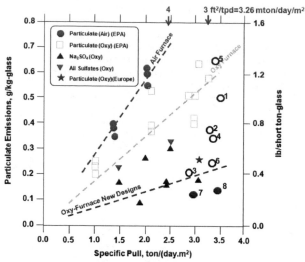

Figure 6. Particulate Emissions from Natural Gas Fired Container Glass Furnaces

In Figure 6, actual particulate emission data collected from several air and oxygen fired glass furnaces are correlated with the specific production rate per unit melter area. The specific production rate was compensated for electric boosting by subtracting the equivalent amount of glass produced by electric boosting. The round and square data points on particulate emissions are based on the U.S. EPA Method 5 and came from three regenerative container glass furnaces with nominal capacities of 68, 135 and 310 metric tons per day, before and after conversion to oxy-fuel firing. The data show sharp increases in particulate emissions per unit glass surface area with increasing specific production rates. This trend is consistent with higher volatilization rates of NaOH expected at high glass surface temperatures and gas velocities when the furnaces operate at high specific production rates. A comparison of these air and oxygen data shows about 20 to 30 % reduction of overall particulate emissions under oxygen firing. In these early oxy-fuel conversions, little design considerations were given to minimize the alkali volatilization and particulate emissions. In later furnace conversions, several modifications were made to the furnace design and the oxy-fuel burners based on detailed CFD studies of alkali volatilization and actual furnace measurements of the concentration of alkali species[3-4]. The design improvements included lower flame velocities, a higher burner elevation and the taller crown[5]. Triangular data points represent sulfates measured by extracting flue gas samples with particles and analyzing by the ICP method. Typically 80 to 90% of "particle emissions" from soda lime furnaces are submicron particles of condensed vapors of sulfates. Particulate emissions were reduced by approximately 50% by improved furnace/burner designs as compared to the earlier oxy-fuel furnace design and the dotted line represents the trend with the new oxy-fuel furnace design. (Note: EPA Method 5 includes H2SO4 as particulates and makes the amount of particulates higher by ~40-100% vs. the European Method)

Particulates measured from air and oxy-fuel fired furnaces equipped with batch/cullet preheters are plotted as round data points with numbers 1-8. All these data points show high specific pull rates of 3 to 3.5 mton/day/m², which demonstrate an important benefit of preheated batch/cullet. Data points 1-4 are from four different air fired regenerative furnaces with dry preheated batch/cullet. Data points 4-6 are from the same end fired furnace[6], but with different batch conditions. With wet batch and no preheating (data point 6) particulates captured above the checker were about 0.23 kg/mton. With dry preheated batch containing fine particles (25% less than 100 μm) the particulates carryover increased sharply to 6.5 kg/mton (data point 5). When fine particles were removed from batch (3% less than 100 μm) particulates carryover were reduced to about 0.33kg/mton (data point 4). These data points confirm higher particulates carryover with dry preheated batch, especially with fine batch particles, when applied in air-fired regenerative furnaces.

Data points 7 and 8 are from two oxy-fuel fired furnaces from the same plant using same batch composition. With wet batch and no preheating (data point 7) particulates at the furnace flue port exit was about 0.11 kg/mton (0.22 lb/ston) at specific pull of 2.9 mton/day/m^2. With dry preheated batch/cullet specific pull was increased to 3.4 mton/day/m^2 and particulates captured increased to 0.14 kg/mton (0.28 lb/ston). Most of the increase in particulates emission, however, are considered to be caused by the increase in specific pull when compared with the slope of the trend line.

A more detailed elemental analysis of captured particulates is presented in Figure 7.

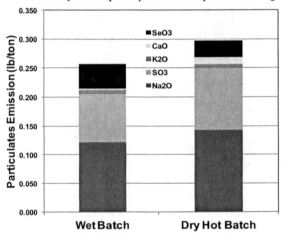

Figure 7. Particulate Emissions from Oxy-fuel Fired Furnaces: Wet vs. Dry Batch

Most of the increase in the overall particulates emission is shown to be due to the increase in volatilization as indicated by the increase in Na2O+SO3 (i.e.Na2SO4). CaO increased from 0.002 lb/ton to 0.012 lb/ton, which was probably caused by dry batch dusting. It corresponds to about 3.5% increase in the particulates emission at the same specific pull. Data also show a benefit of reduced volatilization loss of selenium as the SeO3 emission decreased from 0.005 lb/ton to 0.004 lb/ton. Faster glazing and melting of preheated batch is probably responsible. These results support the view that a properly design oxy-fuel furnace is capable of using dry batch without significantly increasing particulates carry over.

Two of current Praxair offerings for the glass industry further improve the performance of oxy-fuel fired glass furnaces by reducing particulates carry over and volatilization from glassmelt. They are Tall Crown Furnace design and ultra-low momentum DOC WideFlame Burners (DOC-WFB). Tall Crown Furnace design is an evolutionary oxy-fuel furnace and burner design developed from extensive R&D, field tests and commercial operation[6]. The basic design concept is to raise both the side-wall height and the elevation of oxy-fuel burners placed in the side walls, which reduce the furnace gas velocity near the glassmelt surface and the circulation of volatiles to the crown. The DOC WideFlame Burner produces ultra-low momentum flames that reduce volatilization of alkali species from glassmelt surface. The burner also produces a luminous wide flame for better heat transfer and achieves up to 50% NOx reduction compared to conventional oxy-fuel burners[7].

PART 3: BATCH WETTING- HANDLING AND PREHEATING

Matthias Lindig-Nikolaus Sorg GmbH, Germany

The European Union Framework Directive on Eco-Design of Energy-Using Products (Directive 2009/125/EC) 1 establishes a framework to set mandatory ecological requirements for energy-using and energy-related products sold in all 27 Member States. Its scope currently covers more than 40 product groups (such as boilers, light bulbs, TVs and fridges), which are responsible for around 40% of all EU greenhouse gas emissions.

It establishes a list of 10 product groups to be considered in priority for implementing measures in 2009-2011: Amongst the list of product groups are industrial and laboratory furnaces and ovens for all kind of application including for glass. An overall assessment has been carried out comprising the state of the art of the available technology. This may give an indication for how seriously the energy savings issue will be treated by authorities within the next years entering into clear guidelines. The clear demand will be reduction in heat losses through walls and flue gases. Waste heat recovery has to comply with the entire processing and it should go along with efficient and cost effective solutions. Energy consumption reduction is of primary importance.

Batch wetting might be one of those issues which have to be reconsidered regarding benefit and drawbacks. When batch preheating by use of hot furnace flue gas is considered removal of the majority of batch wetting moisture will be essential. The water addition to batch is mostly consumed by the Soda Ash. The Soda Ash incorporates the water reforming from Mono Hydrate to Decca Hydrate. Above 40°C the Decca Hydrate discomposes again to Mono Hydrate releasing the water to the batch. The ingredients Sand and Limestone may combine with the water to form Calcium-Silica- Hydrate below 100°C and causes batch clumping. Due to this, any batch preheating solutions will require less than 3wt% water in batch.

In relation with water in batch the batch preheating has some significant advantage. Wet batch is fed into the furnace and heated up. The water consumes heat for evaporation and for further heat up to combustion space temperature which is above 2/3 of the total heat digested by the water. Preheating batch with hot flue gas helps to release the water already in the heat exchanger. No further heat up of the steam in the furnace takes place.

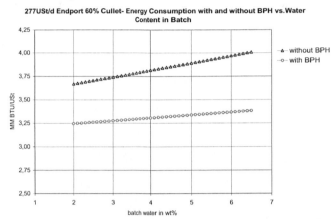

Figure 8. Energy consumption with and without (Batch Pre-Heating) BPH vs. water in batch

Even with low water content in batch and unpredictable variations there is still the risk of clumping in the heat exchanger. The Sorg Company has carried out various investigations on heat exchanger technology which implies a solution against that clumping and plugging. It might be desirable to use mechanical activation in the exchanger preferably in the section where the batch temperature varies between 20 and 80°C. Even with low water content of the batch it is desirable to perform an immediate drying.

The Sorg typical feature, patent applied for, is no coverage of the heat exchanger with batch in order to perform a proper release of the moisture. Since the preheated batch, after leaving the preheater is very volatile, it is highly recommended to work with a completely sealed charging pocket, in order to avoid dusting outside the furnace. Inside the furnace the risk of dusting (carry over) will be even more serious due to chemical attack of the refractory in the superstructure and regenerator chambers. Sorg is recommending to use an enlarged charging pocket which allows radiative heat entering into it and helps early glazing of the batch surface.

Before inserting the batch preheat technology a number of boundary conditions have to be discussed. The water content of the batch dictates how complex the heat exchanger finally should be. The more water the batch contains the more slowly the mixture temperature will rise inside the heat exchanger and larger the portion of the total exchanger where mechanical activation will be required. This goes with a significant increase in investment costs. Since the source of excess water in batch is not due to preparation but due to wet cullet it might be recommended to reconsider the way of storing the cullet.

Another point of view addresses the volatiles and carry over inside the furnace. The enlarged doghouse may help to glaze the batch surface entering the furnace. It does not prevent gaseous release of components or heavy decrepitation during heat up before being converted into melt. This source of volatiles is also not affected by the batch wetting. Decrepitation takes place when grains with strong mechanical stress inside are heated up quickly. The impact will be even worse when dry hot batch is used. The decrepitation may carry off other fine ingredients of the batch into the combustion atmosphere! The decrepitation and the evaporation and final condensation in the flue gas can be the major source of the total dust collected behind the furnace. Investigation of the heat up behavior of the ingredients is recommended. In case of container glass, the main source of evaporation and condensation is the Soda Ash and the Sulphate. The gaseous components react in the flue gas generating Sodium Sulphate. It is recommended to investigate to lowest necessary amount of Sulphate needed for acceptable refining performance.

The process of planning a batch preheater starts with investigating the temperature window allowed for the heat recovery. The heat content at temperature level of the flue gas behind the regenerator and the required temperature level of the filter system inlet are the boundary conditions for the heat exchanger. The boundaries are crucial for the final exchanger design, in particular the inner heat exchanger surface. Calculations have been carried out to show the overall efficiency of the furnace including preheater with different sizes of the regenerator. Since the efficiency of the regenerator is higher compared to the batch heat exchanger it might not be recommended to downsize the air preheating in favor of improved batch preheating. It is self evident that even with enlarged exchanger surface the efficiency of the preheater will not rise by the same magnitude.

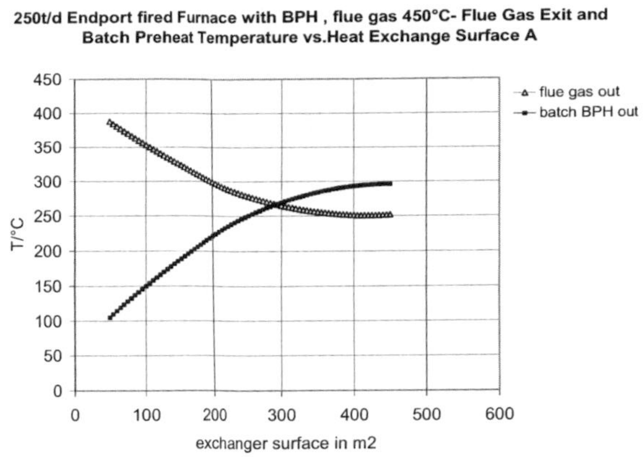

250t/d Endport fired Furnace with BPH , flue gas 450°C- Flue Gas Exit and
Batch Preheat Temperature vs.Heat Exchange Surface A

Figure 9. Flue gas exit and batch Preheat temperature vs Heat exchange surface A

The heat exchange calculation is strongly dependant on the design of the exchanger (residence time of batch, distance between the exchanger members). Enlarged exchanger surface use to comprise also increased pressure drop. Stronger exhaust fans will be required. False air sucking in through leakages may increase the total amount of flue gas. Other preheating technologies may help to improve the efficiency like raining bed heat exchanger. Even those solutions have to be combined with a separate preheating sections prepared for the critical first heat up below 100°C. In every case either green field or retrofit it is recommended to reconsider all boundaries and to calculate the most convenient size of the heat exchanger facility. Optimizing the boundaries, like water reduction in cullet, may be very suitable in order to keep the costs inside reasonable limits.

REFERENCES
1. Lehman, R.L. and Manring, W.H, "Batch Wetting With Water", The Glass Industry, December 1977, pp16-34
2. Mutsaers, L.M., Beerkens, R.G.C., De Waal, H., "Fouling of heat exchanger surfaces by dust from flue gases of glass furnaces, Glastech. Ber. 62 (1989) Nr. 8 pp266-271
3. Kobayashi, H., Wu, K.T. and Richter, W."Numerical Modeling of Alkali Volatilization in Glass Furnaces and Applications for Oxy-Fuel Fired Furnace Design", Glastech. Ber. Glass Sci. Technol. 68 C2 (1995) pp119-127
4. Wu, K. T. and Misra, M. K., "Design Modeling of Glass Furnace Oxy-Fuel Conversion Using Three-Dimensional Combustion Models," 56th Conference on Glass Problems, University of Illinois at Urbana – Champaign, IL, October 24 to 25, 1995.
5. Pörtner, Dirk, "Experiences with an Oxy-Fuel Container Furnace," Glass Industry, May 1999, pp. 25 to 28.
6. Beerkens, R.G.C., "Energy Saving Options for Glass Furnaces and Recovery of Heat from Their Flue Gases", 69th Conference on Glass Problems, Columbus, OH, November 3-4, 2008

Author Index